Branding and Product Design

Branding and Product Design

An Integrated Perspective

MONIKA HESTAD

Routledge
Taylor & Francis Group

LONDON AND NEW YORK

First published in paperback 2024

First published 2013 by Gower Publishing

Published 2016 by Routledge
4 Park Square, Milton Park, Abingdon, Oxon OX14 4RN

and by Routledge
605 Third Avenue, New York, NY 10158

Routledge is an imprint of the Taylor & Francis Group, an informa business

Publisher's Note
The publisher has gone to great lengths to ensure the quality of this reprint but points out that some imperfections in the original copies may be apparent.

Gower Applied Business Research
Our programme provides leaders, practitioners, scholars and researchers with thought provoking, cutting edge books that combine conceptual insights, interdisciplinary rigour and practical relevance in key areas of business and management.

British Library Cataloguing in Publication Data
Hestad, Monika.
    Branding and product design : an integrated perspective.
    1. Product design. 2. Brand name products. 3. Branding (Marketing)
    I. Title
    658.5'752 – dc23

The Library of Congress has cataloged the printed edition as follows:
Hestad, Monika.
    Branding and product design : an integrated perspective / by Monika Hestad.
        p. cm.
    Includes bibliographical references and index.
    ISBN 978-1-4094-4626-2 (hardback : alk. paper) -- ISBN 978-1-4094-4627-9 (ebook)
  1. Brand name products. 2. Product design. 3. Branding (Marketing) I. Title.

    HF5415.1255.H47 2013
    658.8'27 – dc23

                                                            2012033345

ISBN: 978-1-4094-4626-2 (hbk)
ISBN: 978-1-03-283721-5 (pbk)
ISBN: 978-1-315-56987-1 (ebk)

DOI: 10.4324/9781315569871

Contents

List of Figures

List of Tables

About the Author

Dr Monika Hestad is a brand and design strategist and an associate lecturer at Central Saint Martins College of Arts and Design (CSM) in London. As a consultant on behalf of CSM, Dr Hestad has over several years been involved in training leading fashion entrepreneurs and designers in Beijing in how to use creativity to build global brands.

Dr Hestad has more than a decade's experience as a designer, and set up her own design and branding consultancy, Brand Valley Design, in 2003. After relocating to the United Kingdom in 2009, she joined the product strategy firm Plan Strategic and was particularly involved with a leading global company in fast-moving consumer goods markets. She has conducted branding and design work in high-tech industries and service sectors, working with brands including Nokia, Yellow Pages, Innovation Norway, Ibruk, Thrane & Thrane, NokiaSiemens Networks, EB-Elektro, Mars, Jøtul, Stokke, and oral care in Procter & Gamble.

Dr Hestad has also been involved in policy-making, and was part of the Norwegian Government Commission that proposed the Norwegian Higher Education Act 2005.

In addition to regular teaching at CSM, the Oslo School of Architecture and Design and in Beijing, she has given lectures at the Bergen National Academy of the Arts, University of Skövde (Sweden), Oslo University College, Oslo School of Management, Ravensbourne, and taught students at the joint Seoul School of Integrated Sciences and Technologies (aSSIST) and Aalto University MBA course in Korea.

Dr Hestad gained a Master of Industrial Design degree (sivilindustridesigner) from the Oslo School of Architecture and Design in 2003, and also spent time in Paris as an Erasmus Scholar at the École nationale supérieure de création industrielle (ENSCI/Les Ateliers). She received her PhD in Branding and Design from the Oslo School of Architecture and Design in 2008.

Acknowledgements

It is more than a decade since I started asking questions about how the brand is linked with the product, and how I as a designer could embed the idea of the brand in the product design process. Initially, my goal was to develop a tool that could help me transfer the idea of the brand into the product. However, after a bit of research I changed, the world changed, and I now find that the question has become even more relevant more than ten years after I first asked it.

This book is the result of many conversations with key practitioners in the industry, industry partners, as well as friends and family. This has been a long journey, and I am thankful to everyone who has taken time to contribute. The list of acknowledgements describes the long process that led to it being written.

To start with, I would like to thank my partner, Anders Grønli. Without his help this book would not have been possible. Not only has he spent every day encouraging me in my writing, he has also commented on the chapters as they developed. Anders has contributed with editing and included his own thoughts. Parts of the book are also based on his own research, to such an extent that I was tempted to include him as a co-author.

An important part of developing this book has been dialogue with partners in the industry. I have appreciated working with passionate entrepreneurs to learn about their businesses and their needs. This has given me an opportunity to test ideas and theories against reality and to develop a framework that could be relevant for their businesses. Thank you to all. Thanks also to Brand Valley Design Ltd for financing the time it has taken to write this book.

This book would not be possible without the willingness to share from key people I admire in the design community. The material the book is based on developed through years of teaching at the Oslo School of Architecture and Design (AHO) in Norway and later on the MA course in Innovation Management at Central Saint Martins College of Arts and Design (CSM) in London. Without the support of my colleagues at these two institutions I would not have been able to develop the framework the book is based on. Without the opportunity to teach I would not have seen the need for this book, something that has been an important motivation along the journey. I would especially like to thank Nina Bjørnstad, Professor Simon Clatworthy, Professor Halina Dunin-Woyseth, Anne Mellbye, Professor Jan Michl and Professor Rachel Troye for their continuous encouragement. Since before I graduated, Nina has been a helpful partner in discussing this field. She also took time to comment on parts of my book. Professor Jan Michl played an important role as my doctoral supervisor. Once again, his valuable and insightful comments helped me to develop this book further. Professor Halina Dunin-Woyseth is a source of inspiration for young researchers. Her constant encouragement since the start of the project is much appreciated.

At CSM, I would like to thank Dr Jamie Brassett, Maria Ana Bothelo Neves and Rakhi Rajani for providing valuable insights on research and innovation. In writing the book, they have shared their perspectives on the industry. Rakhi has, in particular, given me valuable input to Chapter 7 on innovation and branding. I would also like to thank the students I have been teaching for what they have taught me. Having thirty critical minds to question, comment and add their own stories to the material I am presenting has been invaluable to my writing. Seeing the areas my students are interested in has been important in understanding gaps in the industry as well as in the literature.

This book follows up my doctoral dissertation, *Den kommersielle formen* (Hestad 2008). Although some time has passed since I conducted my doctoral research at AHO, the cases I had selected were still relevant for this book and provided an important framework to build on. I would therefore like to thank all the people I interviewed for my thesis. This includes the people at Stokke, particularly Hilde Angelfoss, who also provided comments when writing this book. It also includes contributors at Jordan Dental Care, especially Michelle Wentworth and Geir Hellerud, and the designer of one of their toothbrushes, Geir Øxseth, as well as the people at Ringnes (Carlsberg), the Scandinavian Design Group and Frank Design. I would also like to thank Design Bridge, Tinhorse and P13 Innovation for taking the time to see me for an interview. It is eight years since I interviewed them, and they probably do not even remember that I started this process – finally, here is the result.

Since the start of my research career the Nordic design research network Nordcode and the Norwegian network DesignDialog have been important arenas for scholarly discussions. Dr Ingvild Digranes at Oslo and Akershus University College and Dr Martina Keitsch at the Norwegian University of Science and Technology (NTNU) have been valuable discussion partners.

Moving to London and working at Plan Strategic was crucial in seeing the relevance and importance of the questions I asked. I am sending thanks to my former colleagues at Plan for giving me insights into the English design community. Particular thanks go to the founder of Plan, Kevin McCullagh, for sharing his perspectives on experience design and branding in the second round of interviews I conducted in the United Kingdom. I would also like to thank Sanjeev Davidson for sharing his knowledge about branding in the fashion industry, and Michael Chen and David Wang at Sigen in Beijing for giving me a glimpse into the exciting fashion brands that are in development in China.

In the final stage of developing the book, I am grateful that Joanna Brassett at INTO, Damian Mycroft at Hewlett-Packard, Paul Marchant at Transport for London and Cathrine Movold at Making Waves, who gave of their time and were as open as they were in sharing their insights from their own work. Their contribution was important to clarify my own thinking and to test my ideas. A special thank you to Paul and Cathrine, who also commented on other parts of my book and encouraged my writing. I would like to thank Julia Cullen, Paul Sturrock and Jonas Altman in London and Dr Toni Matti Karjalainen at Aalto University in Finland for taking the time to comment and read parts of my book. I also appreciate Silvia Rigoni's encouragement and enthusiasm in the final stages of finishing the book. My appreciation also goes to Gower Publishing, and especially to Martin West and Emily Ruskell for believing in a first-time author.

Friends and family have been patient throughout the journey that led me to this book. Thanks to all of you for being tolerant and supportive. My local café in Deptford, The Hoy, and Claire McCauley and later our favourite Highgate pub, The Bull, have provided excellent atmosphere and food during many hours of writing this book. Finally, I would especially like to thank my mother, Karin Brandal.

Thank you all! This book reflects how far my thinking has come based on the input from you. I hope that you enjoy reading it, and I am looking forward to continuing our conversations.

Introduction

Product design and branding are two activities that arose with the Industrial Revolution. Moving into a knowledge economy, both activities have become increasingly sophisticated, and are adding more value to products. However, product design and branding have also been recognised as two tools that, when misused, can create a fragmented and weaker appeal for the company or its output. This is a particular risk when brand strategists try to adapt existing products to new ideas about the brand. An example of this would be a brand strategist who defines a new brand personality without understanding the feelings and use that consumers have connected to the product. If the product is well established and closely tied to the brand, and the change is made without an understanding of what the product is about or why consumers liked it in the first place, the brand strategist is risking the very foundations of the brand. At the same time the risk is just as great if designers suggest a new style based on their own preferences in a way that disregards the heritage and context of the brand.

The misuse of design and branding must be seen in parallel with the argument put forward by the brand theorist Michael Beverland that brands like Apple and Harley-Davidson have an enormous appeal because they appear 'less interested' in marketing than others, while other companies have in contrast 'invested enormous sums into marketing and design' and still lost public appeal (Beverland 2009: 1). Still, looking at Apple and Harley-Davidson, it is evident that they are devoting considerable amounts of money to both branding and design. Their appearance, however, feels less 'added' and more coherent with their reason for being. The perspective of this book is that both design and branding should be embedded in what the company is doing. Most importantly, this integrated perspective needs to be driven by values and a reason for being rather than creating a superficial position in the market that the company can defend. In the branding activity, the product design can serve as an interactive dialogue, where past meets the future of what the brand means for consumers. This book will examine how the product can fulfil an important role in building the brand – and in particular how the product design process can contribute to building meaningful and relevant brands.

Change of Mindset

Being involved in meaningful activities and a company that is driven by values is more important than ever before. In part this is because the development of our societies has allowed us to be discerning and discard 'less meaningful' activities. Another factor is that there has been a fundamental change in how people see companies' roles in society, and perhaps how companies define their own roles.

For decades Milton Friedman's (1970) position that the only social responsibility a company has is to make a profit has had an influence on companies. However, a much-praised article by Harvard University researchers Michael E. Porter and Mark R. Kramer in the *Harvard Business Review* (2006) suggested that this was about to change. The focus on monetary profit has led to an emphasis on short-term gain to the detriment of wider society. The article by Porter and Kramer introduced the concept of 'creating shared value' (see Porter and Kramer 2006; Porter and Kramer 2011). In short, this concept suggests that the company will enhance its competitive advantage while 'simultaneously advancing the economic and social conditions in the community between societal and economic progress' (Porter and Kramer 2011). This is relevant to the discussion of design and branding presented in this book, as Porter and Kramer's perspective enforces a focus on values that lie beyond financial profit. This is a long-term perspective where the company is driven by social and economic values, and is in dynamic interplay with society.

The redefinition of companies' role in society will lead to companies being increasingly engaged in meaningful activities. This is a timely perspective, as present and future employees will tend to seek companies that offer them meaningful tasks. The American sociologist and economist Richard Florida pointed to the 'rise of the creative class' in a widely read book that came out in 2002. The creative class is a group of people where the job is important, but only if it is a meaningful activity. Members of this class are educated and share a common set of values, such as creativity, individuality, merit and difference. The people in this class do not see themselves as a class – it is the shared mentality that makes them a coherent group. The mentality that drives their work is that they should not only be paid well by their employers, but should also have like-minded colleagues who share their values.

Another interesting development that goes hand in hand with this changed mentality is the rise of mass creativity that has been made possible by new technology. Creativity has gained new access to easy and low-volume manufacturing, as well as new, democratic promotion channels. The last decade saw change because of the 'You are what you share' mentality described by Charles Leadbeater in his book *We-think* (2009). *We-think* examines how the masses have been able to create a new form of identity over the past decade through the opening up of social media spaces and easy access to blogging.

The Economist's 'The Ideas Economy' blog tells the story of this focus on creativity: '*The Economist* believes that human progress relies on the advancement of good ideas' (*The Economist* 2012). This interest of a leading news magazine such as *The Economist* in creativity and design indicates that the macro picture is changing. The rise of the creative class and the increased role creativity plays in the economy signals a change in people's attitude to the workplace that is value-oriented, not just in terms of financial value, but also in terms of 'being meaningful'.

These changes in society are also altering attitudes to brands and products. The emphasis on creativity has also led to increased knowledge about the results of creative activities. People in general have a better understanding of both design and branding. This applies not only to the 30 per cent Richard Florida believes can be counted in the creative class – people in general have a better understanding of branding. With the increased use of software programs at an early age, young people are used to being creative and playing around with typefaces and becoming skilful in design. In branding, the emphasis used to be on creating a constructed identity that was promoted through all the touch-points the company has with its customers. This is no longer sufficient, as consumers tend to 'poke holes' in shallow identities (Holt 2002). To build a brand with substance, it is necessary to consider all of the company's activities, and also how people outside the company relate to it.

Products' Role in Branding

These days, it is not only business units, branding departments or marketing departments that are in charge of telling the story of a brand: 'A brand emerges when various "authors" tell stories that involve the brand' (Holt 2004: 3). These stories are told by the managers, the design team, the marketing team, the sales team and all the other members of the organisation. The stories of the brand are also told by the customers and the cultural industry, and the Internet has opened up multiple channels that people can use to communicate their own versions of the brand. This has changed the game, so the product and how people are experiencing it has become far more important.

In brand-building, the product is portrayed as one of many touch-points that communicate what the brand is about (Wheeler 2006). The product becomes a very important touch-point because it is the key physical representation of the brand, making intangible concepts real for the customer. The product can therefore be understood as the substance in the brand story, and what prevents the brand being purely 'smoke and mirrors' (Aaker and Joachimsthaler 2000), becoming instead the substance the product will also need in order to offer value. Brand theorist Michael Beverland has investigated what makes brands iconic. He found that consumers respond to product-oriented brands 'because they make substantive commitments to quality instead of trying to sell poorly performing products with advertising campaigns' (Beverland 2009: 109).

The product is important in telling the story about the brand by delivering the brand promise. The aim of this book is to explore how the design process can serve an important function in telling the story of the brand. The design process gives the company an opportunity to maintain the strength of the brand, to reposition the brand and to establish it. But in order to do so, there is a need to view the product design process from a strategic perspective that is driven by a desire to create shared value, both for the company and society.

The Strategic Importance of Design

The strategic importance of design has been recognised for some time, and has been a 'hot potato' in the design community. In designer brands, like the office furniture maker Herman Miller, designers have traditionally played a prominent role – not only in product development, but in the organisation as a whole (Beckwith 2004).

In 1907 the architect Peter Behrens was employed in a strategic position by Emil Rathenau, the founder and President of Allgemeine Elektricitäts-Gesellschaft (AEG), a major German electrical equipment company. Behrens was hired to design not only products and buildings, but also the 'corporate identity and print advertising' (Vogel 2010: 6).

Another classic example of focusing on the role of the designer is General Motors (GM) under Alfred P. Sloan's legendary leadership. The industrial designer Harley Earl had created several successful designs, and had gradually been given greater responsibilities. In September 1940 Sloan promoted Earl to a Vice President position at GM, which Sloan believed was the first time a designer had been given a post at this strategic, executive committee level in a major company (Sloan 1986: 277).

As mentioned above, designers in companies have had success through strategic roles where they have been able to plan activities for the potential future of the brand. Nevertheless, the majority of designers find themselves in a tactical position within companies, where their role is to design according to specifications set by non-designers. Designers also tend to find themselves coming in at the very end of the product development process, when it is a question of 'adding value' by creating a style that corresponds with current trends. However, this business model is not viable when creating brands with substance.

There can be many reasons why designers continue to find themselves regarded as merely an additional marketing cost in many companies, despite designers having been very successful in some well-known companies. One reason is the lack of commercial and strategic training of designers in many design schools. Historically, product designers have been positioned between engineers and artists (Pye 1978).

There is a gradual development to increase the business perspective that exist beyond technology. Studies from Finland show that designers from the late 1980s increasingly have taken on a role of a coordinator, both including an end-user perspective and perspectives from units within the company such as marketing and engineering (Valtonen 2007: 296). The Finnish design researcher Anna Valtonen (2007: 304) points to how this in the past two decades has been developed further to see designers push innovation and belong to the team that creates the corporate vision. This positive development in businesses still leaves a question of whether design schools are equipping their students to take on such roles, coordinating design, business and technology.

In the complex context in which the designer operates, the designed object needs to embody the perspectives of engineers, artists, marketers and researchers. The Global Head of Consumer Experiences at toy manufacturers Lego, Cecilia Weckström, emphasises the importance of learning the language of managers, and not using what she referred to as 'designease' (Weckström 2011). If designers want to take on a strategic role, they also need to understand the appropriate language, and to conceive of the designed product as part of a wider context.

Where Product Design Meets Branding

Branding and product design are presently two separate professional disciplines. When looking into how these two can be combined, every element of the product design process and the brand-building process is relevant. The challenge of writing this book was therefore to include enough information about these fields to understand where they connect. There are many books that describe the design process in detail, and how designers work in the different stages from research to implementing their ideas in the product and industrialisation. Likewise, there are many books about how to establish a brand, and how to build a brand in an organisation or in the corporate identity. Building on theories from both of these fields the aim of this book is to convey a holistic understanding of what they involve and how they can be connected.

The intended audience for this book includes those who would like to learn more about the strategic dimension of product development, focusing on the role products can play in building brands. It is written for people who are interested in the 'what', 'why' and 'how'. While the primary audience will be designers, it will also be useful for marketers and managers who are interested in understanding how the product is significant in building a brand with substance.

The book will explore some of the key issues that should be raised in viewing the product development process as an important aspect of building the brand. It draws on key literature within the branding and design fields, and references will point to further reading. It is also based on original research conducted as part of my doctoral work (Hestad 2008). In addition, it includes interviews with a number of people within industry, with backgrounds ranging from design to marketing and management. Each chapter has been written so that it can be read both individually and as an element within a continuous book.

Chapter 1 explains what branding is about and discusses the role of the product in six different strategies. The chapter culminates in a diagram showing six different brand stories and how the product is involved in each of them.

In order to understand both design and branding – and the state of these fields today – it is important to establish a historical perspective on how these professional areas have developed. Chapter 2 offers an introduction to brand history in order to learn how the relationship between products and brands has changed. It also provides a short introduction to design history from the perspective of building brands.

A product with a story is crucial in building a brand. Chapter 3 therefore explores how consumers experience products, and how products play their part in telling various brand stories.

An important aspect of building a brand successfully is to understand the full context of the product within the design process. Chapter 4 presents a three-layered model of how the design process can contribute to building a culturally relevant and dynamic brand.

How the product design process functions within the branding process is a management question. Chapter 5 analyses how organisations can maintain a knowledge base, who should design a product, and how to translate the brand DNA into something that can be used as the basis for design.

Chapter 6 explores the importance of informed decision-making, in particular the vital role of research and how the information it provides can be used to define the brand.

Finding the right balance between new ideas and maintaining what is already recognised is a key consideration in using the product as part of the branding exercise. Chapter 7 discusses the relationship between the innovation process and the brand, and in particular how innovation can be a powerful tool to keep the brand up-to-date and relevant.

To summarise: this book will provide background for designers, marketers and managers who want to find better ways to use both products and brands in a joint approach to achieve meaningful interaction with customers.

1

Strategy: The Brand and the Product

From all the money we invest in building a brand, it is through the product we have the best chance to communicate what the brand is about. The product is what the consumer sees most of. We can leave out advertising and other activities to promote the brand, but the products have to represent the brand Jordan. The product must give a signal, it has to attract you into buying Jordan, and the product has to represent the brand. It has to be good in use and give the user a good experience because it represents the impression you have of the brand. We see the product as 'the person' representing the brand.

Michelle Wentworth, Category Development Manager,
Jordan Dental Products

This first chapter describes six brand-building strategies, as well as the integral role of the product in the branding process. The aim of this chapter is to give an overview of the various strategies companies are using in building their brands, and how their products are closely linked to these.

Why Brands and Not Products?

THE NEED TO BUILD BRANDS

In the early days of the Industrial Revolution the role of companies was to produce products. One company above all others took the lead in changing the game of the industry for ever. The English ceramic company Wedgwood started to build what we today recognise as a brand. Wedgwood was founded in 1759 by Josiah Wedgwood. The innovative and forward-thinking founder invented many of the products that continue to be manufactured to this day. He also earned a place in history by being the father of modern marketing. Bringing several innovations to market, Wedgwood witnessed that his products were being copied by competitors (McKendrick et al. 1982). Since his competitors did not have to fund costly research, they could sell his inventions more cheaply than Wedgwood could. This jeopardised the products' profitability. He came up with the idea of associating his products with the aristocracy, and through this embedded the flair of their luxurious lifestyle in the products. His approach to countering copycats was therefore to court the aristocracy.

It is more than two centuries since Josiah Wedgwood used what today would be described as 'celebrity endorsements' of his products – a powerful brand-building strategy. The reason for building a brand rather than simply producing products remains the same as it was more than two centuries ago. Product attributes and functionality are easy to copy. However, the reasons for building brands today are not only to enable companies to protect their innovations, but because customers and retailers have established a liking for brands rather than nameless products.

Brands help consumers to make decisions. Consumers recognise brands and buy them because they promise to fulfil a need, due to recommendations or based on earlier experiences with the brand. Consumers also buy brands because the brand story connects with them emotionally, it offers them a self-expressive benefit, or they find the brands relevant in a certain cultural context. For entrepreneurs, building a brand gives an opportunity to build a legacy, to be recognised and to attract the best employees who would like to be part of the culture that is building the brand. Therefore, many companies find that manufacturing products is not enough to compete in the market – and more importantly, mere manufacturing of products is not satisfying on a social level.

WHAT IS A BRAND?

Most people working within design and branding have a good idea of what a brand is. When asking an audience of people with design, marketing and management backgrounds what a brand is, we can easily get thirty different answers. People might describe the brand as 'an idea the customer has of their company', 'the dream of the founder' or 'an emotional feeling that the customer gets from owning a product'. These are all good explanations of what a brand is about.

There are three factors that are important to what defines a brand (see Figure 1.1). First, there needs to be a desire to perform a certain meaning, a philosophy or a vision. To simplify what this is, in this book we will call it _the brand story_. This can be a story that the company decides to communicate, or as we will see later, it can also be people on the street who shape what the brand story is all about. Second, there needs to be something that represents the story. This can be the products themselves, but it can also be the name of the products, a logo or promotion activities to consumers. In the case of Wedgwood, the products were given names with aristocratic associations, and in this way the name encapsulated the 'flair of the aristocrats'. The name becomes _the mediator_ that makes people think about the brand story. Third, there also needs to be someone who interprets the mediator to represent the brand story – _the interpreter_. To qualify as 'a brand', the same story needs to be recognised by a group of people.

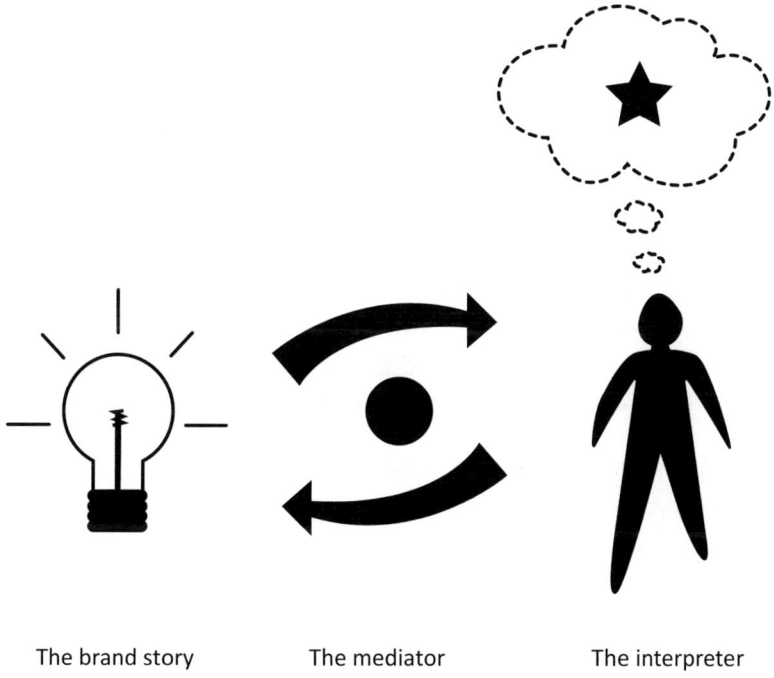

The brand story   The mediator   The interpreter

Figure 1.1   The brand story, the mediator and the interpreter

The true story of the brand is not necessarily the meaning the company intended, but rather what people feel when they think about the brand. A brand is not a fixed entity. The relationship between the brand story, the mediator and the people interpreting and telling new stories about the brand is a dynamic one. This relationship evolves based on each and every activity of a company and everyone involved in the different touch-points of a brand. If there is a difference between what the company would like to communicate and how people interpret the story, there is a brand gap. For the company, it will be important to identify these brand gaps, as over time they may create a fragmented identity which people do not associate as a brand. Product design can play a significant role in strengthening the brand, as well as repositioning the brand.

THE RELATIONSHIP BETWEEN PEOPLE, THE PRODUCT AND THE BRAND

There has been a shift in how product designers are working. Branding and product design have typically been seen as two separate activities, while today designers will have to incorporate brand awareness into the development process (Mycroft 2011). Designers are also involved in planning the customer experience of the brand beyond the product itself. When it comes to services, designers are creating what can be seen as touch-points between the organisation and the consumer, and strive to establish an optimal customer experience. In these touch-points the designer designs the whole experience the branded service provides. In mobile phones, for example, the holistic integration of the interface and the tangible object is crucial. The product design process is therefore seen as one of many activities that are undertaken to build the brand. The relationship between people and the brand can include a product as the mediator of the brand story, but this is not always the case.

According to Kevin Roberts, CEO of the advertising giant Saatchi & Saatchi, the relationship between people and brands is defined by the intensity with which people love and respect brands (Roberts 2004). A few of the strongest brands, such as Nike or Apple, have managed to win a unique position in the market because they have earned both the love and the respect of their customers. While the choice of strategy that will be right for a particular company is outside of the scope of this book, it is worth mentioning that the brands Saatchi & Saatchi has identified as 'Lovemarks' all have a strong product or service story that is actively part of building the brand. These brands have all earned love and respect from their consumers.

Others, such as Hello Kitty, are recognised internationally as strong brands that many young people have an emotional connection with. This brand has a strong position in the market and is deeply loved. Some parents may well be puzzled when they find that their three-year-old daughter has developed a preference for this particular brand without their knowledge. However, it is almost impossible to avoid.

When the company Sanrio first launched a product with the Hello Kitty figure there was no defined strategy to establish this as a brand (Walker 2009), nor was there a cartoon or an animation that could communicate the story of Hello Kitty. Consumers responded emotionally to the figure, which created an opportunity for Sanrio to exploit. Product design plays a less important role in building this brand as these are generic products, but the products serve as billboards for the Hello Kitty figure. The figure is a powerful communicator of this brand that young people connect with. In the Hello Kitty brand, the figure is the brand mediator. It is reprinted on a number of products that without the figure would be generic products with little distinct profile.

DEFINING THE BRAND STRATEGY

A strategy is about shaping the future (McKeown 2012). It represent the overarching aim that the organisation has defined for where it would like to be and why this is the right place for it to be. Creating a strategy involves both analytical capabilities and imagination as the strategist will need to foresee potential future opportunities as well as threats. The company will most likely have strategies for its organisation, for communication and for the product. The strategist Michael Porter (1996) recognises that the company's marketing position occasionally serves as the company strategy. He uses IKEA as an example of targeting young people who would like to have access to style at a low cost. Porter's article was written in 1996, and almost two decades later it is evident that this marketing concept has proved to be a competitive strategy in the global economy. Seeing the brand strategy as an integrated part of the company strategy is also what this book builds on.

A brand strategy is a plan or a policy for what the company would like stakeholders to recognise the brand as. A brand strategy goes beyond a communication strategy, since all activity in the organisation is part of building the brand. The brand strategy needs to consider what the brand story should be, and why this is the right story. It will always need to be dynamic and open to change according to the circumstances.

Key Questions When Defining the Brand Strategy

- Who are you? – The brand story
- Who/what would you like to become?
- What is your desired value proposition? – Experience of the brand/ benefits offered
- What capabilities and traits do you have that make this goal achievable?
- What is the context? – Trends/SWOT analysis
- Who is this for? – Audience/customers
- Who else are out there? – Competitors
- For whom will this have consequences? – External and internal stakeholder analysis
- Which concepts, symbols or references should represent the brand?
- What is the role of design in building the brand?
- How will you measure the success?

Key Questions When Carrying Out the Brand Strategy

- How will you achieve this? – Product, promotion, channels and so on
- What resources will be needed?
- What are the important milestones?

Depending on the context in which the brand exists, the product will have a different role in building or performing the brand. In the Hello Kitty brand the product did not play any role at all, while in the Nike brand the product plays a far more important role in performing the brand story. However, a strategy should not include detailed plans for how to implement the strategy in the organisation or products. The focus at this stage should be on identifying the key elements that are important to consider when developing the strategy. The implementation of the strategy in the design of the product will take place on a tactical level. How well the implementation is carried out will determine whether consumers recognise the product as a factor in building the brand. The brand strategy will need to be implemented in all touch-points to create a coherent brand (Figure 1.2). Therefore, the product will be only one of many key elements that contribute to communicating what the brand is about.

Figure 1.2   The product as one of many touch-points
Based on Wheeler (2006).

How the brand strategy is implemented in the organisation, in the product and in marketing will be important determinants for how consumers will recognise the brand. This is all part of a dynamic process (see Figure 1.3). The brand is a result of the activity in the organisation, the product, the promotion and how consumers recognise and interact with the brand. This interaction is dynamic. The end of one interaction is the starting point of the next.

There are various strategies for how to build a brand, and the product's role in building the story of the brand will differ accordingly. The following six brand stories will examine the role of the product in performing the brand story.

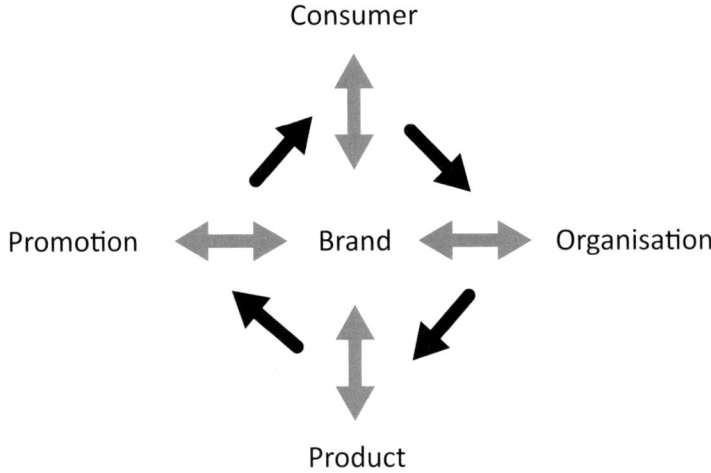

Figure 1.3   Brand as a result of a dynamic interplay

Six Brand Stories and the Role of the Product

FUNCTIONAL BENEFIT: GAINING PEOPLE'S TRUST AND RESPECT

In the fast-moving consumer goods (FMCG) categories there has been a focus on occupying a niche, and establishing either functional or emotional ownership of a message within this niche. Many FMCG products have managed to gain strong positions in the marketplace because they 'own' a functional benefit. A brand may have functional ownership if people believe what the brand promises through a specific functional benefit. This is the case with many of the brands belonging to the British-Dutch conglomerate Unilever, for example. Its brands have for years taught its customers that they are the best in the category. The fear of ruining one's clothes by selecting another washing powder is in many cases greater than the extra cost the customer might have to pay for what is viewed as the premier brand.

By constantly renewing the story in advertisements that state how excellent the company's brand is in providing this functional benefit, Unilever maintains its market position. As long as the brand also manages to deliver on this promise, and constantly innovates to stay ahead of the competition, it will keep this leading market position. Functional ownership is often linked with product performance. The product will be important in capturing the brand promise. Therefore, the company will have to keep innovating its products so that they always perform better than, or at least on par with, its competitors.

EMOTIONAL BENEFIT: GAINING PEOPLE'S LOVE

An approach that focuses solely on ownership of functional performance is seen as outdated in the brand community, as brands are perceived as being emotional entities (Gobé 2001). The emotional dimension will be important for people to recognise a brand. Most brands entail both functional and emotional benefits for the consumers (Figure 1.4). The balance between functional and emotional benefits differs according to the context.

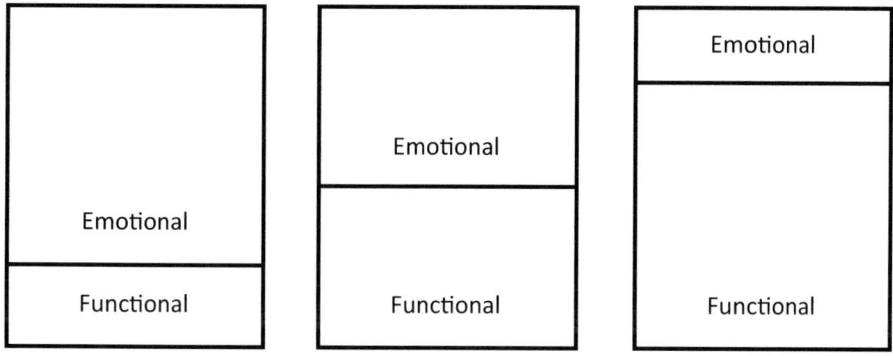

Figure 1.4   The balance between functional and emotional benefits

The oral hygiene brand Oral-B promises a professional deep-clean feeling. This is a brand that is largely based on functional performance. However, underlying this message there is an emotional need. The consumer needs to feel confident or safe. FMCG brands that are perceived as performing better might have this ownership simply because they are the ones we grew up with. This nostalgia adds an emotional layer to our experience with the product.

In the chocolate industry it can be harder to defend rationally that one product is 'performing' better than others. Companies in the chocolate industry therefore often seek emotional ownership of a message. There are a variety of emotional values that the chocolate industry taps into, such as joy, experience and fun. An important dimension in how the chocolate industry will perform its brand stories is to evoke desire and lust, and thus appeal to hedonistic consumers – consumers driven by their own desire to be fulfilled (Blindheim 2004).

The emotional benefit that chocolate brands convey can vary. Gifting is important in the chocolate industry, and many brands, such as Cadbury's Roses, seem to have taken a position as a 'token gift'. Luxury chocolates often claim ownership of a myth or a legend, and promise the ultimate indulgence. The Belgian chocolate brand Godiva is such an example. Lady Godiva is a legend about the beautiful, generous and dedicated Lady Godiva, who was challenged by her husband to ride naked through Coventry. This legend inspired Joseph Draps when he established Godiva Chocolatier in Brussels in 1926: 'He sought a name that embodied the timeless qualities of passion, style, sensuality and modern boldness' (Godiva 2011). The experience of eating the chocolate could be another brand story. The chocolate egg Kinder Surprise from Italian maker Ferrero promises you several things in one chocolate. It promises the delicious combination of brown and white milk chocolate, but also the experience of opening the egg and revealing a little secret toy just for you to play with.

The four-finger chocolate wafer Kit Kat has managed to occupy the 'break', thanks to a coherent communication campaign since 1937. In 1958 Rowntree launched the tagline 'Have a break – Have a Kit Kat' in its first TV commercial. The brand's creator, Rowntree itself, was responsible for this. Nestlé, which took over Rowntree in 1988, then continued the same coherent brand communication.

In a brand story with strong emotional ownership, the product may play an important role even though the product is not directly linked with the brand story. With Godiva, the name of the chocolate comes from a legend – the brand promise is luxurious indulgence. This feeling of luxury is communicated through the packaging of the product, and the product itself is part of fulfilling the experience. The brand story does not come from the product itself – it is captured in the wider context of the product.

In Nestlé Kit Kat, the story is also related to a user scenario. Many people like to eat chocolate during their breaks. By identifying this as an opportunity, Kit Kat has managed to occupy a very important niche in the chocolate industry. The brand story has become very strong as it is linked with a consumer ritual –

breaking the chocolate – and a time of day when people enjoy a piece of chocolate. This link is reinforced in commercials, with close-up shots of someone breaking a piece of the chocolate. The communication of the tag line is also aligned with the product.

Both Godiva and Kit Kat have stories about their brands that are mainly told through touch-points other than the products. The product itself plays an important role in releasing the self-indulgence or luxury feeling. In the chocolate industry, the taste, smell, feel and even the sound of the products will play an important role. Once a good product is established and consumers recognise the brand, the product that belongs to that brand will only go through minor changes. The product will fulfil needs such as indulgence or comfort, and thereby play a role in releasing the brand promise.

SELF-EXPRESSION: HELPING PEOPLE TO EXPRESS THEIR IDENTITIES

Products that are seen as 'fashionable' often have characteristics that consumers would like to be identified with. In these, the brand story will have a self-expressive benefit for people (Aaker 1996). This means that consumers use the products from these brands to express their own identities.

Products that are about mobility are often linked with people's identity. The car is one such product, and so are mobile phones and computers. These used to be functional devices, but can now be seen as fashion objects, both in the understanding that they are about style, and that they are changing according to what is trendy. Even some FMCG products are shifting to focus on expressing identity. The Norwegian bottled water brand Voss, with its stylish, sleek and pure cylindrical bottle with references to the perfume category, has gained a position in the 'fine water' product category. According to a waitress the *New York Times* reporter John Kifner (2002) interviewed, the singer Madonna asked to only have Voss water in her hotel room. Stories like that create a strong identity around the product.

In building a self-expressive brand the brand story could represent a personality, or to achieve a strong emotional connection with the consumer is to build the brand as an archetype (Mark and Pearson 2001). In the fashion industry the founder or chief designer's public personality traits, values and beliefs becomes the brand story. The English fashion brand Paul Smith could be described as 'British classics with a twist'. Sir Paul Smith has given his name to the brand, and is also the chief designer. His personality is embedded in his company's activities. 'Owning' a certain personality means that the brand also needs to have credibility in expressing this. Building the brand so closely to a key person in the company is one approach to gain this credibility.

Brands that have a 'self-expressive' purpose often change to stay up-to-date. The products are then in a constant state of change, and are of key importance both in communicating and releasing the brand promise. In fashion there is a constant drive to change styles every season, and many brands also release mid-season products. However, brands within the fashion industry have a strong identity while at the same time retaining the basic style of their products.

EXPERIENTIAL BENEFITS: FULFILLING PEOPLES' DREAMS

Seeing a brand that is offering an experience as a vehicle to fulfilling dreams has become an increasingly important strategy across all brands. However, for some brands offering the experience will be the core intent of the brand. These are brands where the essence of what is delivered is seldom a physical product, but an opportunity to engage with a story or to take part in an event or an activity. Examples of the output of such brands can be games, movies, plays, sports, museums or other activities. What is offered is to take part in an experience.

A globally leading experience brand is The Walt Disney Company. Their stories are communicated through multiple platforms. They offer their consumers a part in their magical word of imagination. To engage with their story Disney has also established the Walt Disney Parks and Resorts, including theme parks and resorts across the world. The strength in their brand has also allowed them to establish English-language schools for children in China (*The Economist* 2010).

Experience brands are not primarily about physical products, but products will play an important role and be part of many touch-points where consumers can engage with the story. The products could be artifacts that let consumer dress up as their favorite character, gadgets or dolls. Visiting a Disney theme park there will be a range of products that let the consumer engage with the brand from the entrance, to the mascots, to artifacts and gadgets. These will all be part of telling the a story of Disney. The 'Mickey Mouse Watch' was launched in 1933 by what is now Timex (Miller 2010), in a very early example of licensing an experience brand to a physical product. The Walt Disney Company represent a collection of stories that the consumer could engage with. Each single story has a narrative that the physical products will be part of and the products will be dictated both by the narrative in the story, as well as what makes the story a Disney story.

CULTURAL CREDIBILITY: PERFORMING A CULTURAL MYTH

A few brands manage to move beyond others, and could be described as 'iconic brands' (Holt 2004). An iconic brand holds a certain position in society. It holds a special value for people even though they may not be customers of the brand. In 1960 the British shoe manufacturers R. Griggs Group bought the shoe and brand Dr Martens from its German creators, Dr Klaus Märtens and Dr Herbert Funck. The shoe gained enormous appeal among London's sub-cultures, such as the Punk movement. Through being embraced by people with strong identities and value sets, the shoe became a symbol – and an iconic brand was created.

According to Professor Douglas Holt at the Saïd Business School in Oxford, iconic brands have gained their status because of their cultural relevance. The brands hold a special status because they are part of performing what could be understood as a cultural myth (Holt 2004), such as the freedom of being outside the establishment. Harley-Davidson is one brand that has credibility in performing such myths in its story-telling.

Iconic brands have a status among people that is beyond the emotional or self-expressive brand story. They represent something that is different from others in the market, and have earned credibility which over time gives them the status of being iconic brands. In the case of Dr Martens, it appears it was the users of the products, not the company R. Griggs, who constructed the symbolic meaning that is associated with the shoes. This adds credibility to the products that goes beyond what the company could create itself.

In Dr Martens and Harley-Davidson, the products become symbols that represent the brand story. The products are often seen as iconic products. The products are innovative or new to the customer when they are released, but later they remain more or less unchanged, and new product releases are designed to resemble previous models. In these, the development of new products needs to be carefully executed within strict design guidelines. The product has integrity because it is iconic, and iconic products should therefore only be improved with great care and caution.

CREATIVE CREDIBILITY: FACILITATING ACTIVITIES

Today, a new category of brand intent is emerging. The brand story is largely driven by people themselves. A company provides a platform and a process, but people engaging with the process create an important part of the brand story, while the company behind the brand facilitates people's creativity and sharing activities. Facebook has managed to establish itself as a global brand in a very short period. The service was introduced at a time when individuality, sharing and creativity were part of the key trends and motivations. Facebook meets all these criteria.

In short, Facebook allows the user to create a profile and connect with others who also have profiles on Facebook. The brand itself is a provider that makes building these profiles and connecting them a simple process. It provides users with a set of tools they can use to construct their own identities. These tools encompass uploading images, personalising news feeds and playing games. This makes it a powerful tool to enable people to keep track of friends and connections, to build identity and to explore what others are doing.

Facebook is open to companies. The French fashion house Louis Vuitton invited consumers around the world to take part in its product launch. Instead of presenting its new products behind closed doors in Paris, Milan or London, Louis Vuitton launched its 2010 autumn collection on Facebook, with 33,100 users attending the event and sharing the experience of watching the show. In this way, Facebook can become a platform for people, institutions, charities and commercial brands to express their identity and creativity.

The editor of *Wired* magazine, Chris Anderson, introduced the term 'free economy' to describe the economy Facebook is part of (Anderson 2009). In this economy consumers do not pay for the services or the products they are using with money. They 'pay' for these services with their time. Facebook is a 'free' service for users. Facebook earns money by selling customised advertisements based on information users provide about themselves. The activity all these actors create and the number of people engaging on Facebook are important assets for the company.

Both new and established brands are looking into how they can facilitate people's creativity. The American sportswear manufacturer Nike was a pioneer in developing online customisation services for its customers.

Figure 1.5   The artist Geek&Poke's interpretation of the new free model
Creative Commons licence, Attribution-ShareAlike 3.0.

The Italian fashion brand Benetton launched 'me-time' in 2010, and invited people to be part of creating their own campaigns. People could upload pictures of themselves onto Benetton's website and become models for the brand. Brands like Nike and Benetton became well known before it made sense for people to use the brands as platforms for building their creativity. The product played an important role in building people's trust, and offered them the opportunity to express their identity. The next step was to enable people to play with the product in places offered by the brand.

The company delivers the platform and the process that people take part in, while it is the users who deliver the product. The experience of creating and sharing the product is the experience of the brand. The product and the co-creation of the product are integrated into the brand story and the communication.

COMBINATION OF INTENTS

The different brand stories and the role of products in communicating these can be summed up as in Table 1.1. All of these strategies have proven successful for some companies. Which brand strategy the company adopts and how the product plays a role will depend on many factors, such as leadership, time and what is fashionable.

Table 1.1   Brand intent and the role of the product

Brand intent	Example	Story	Product's role in performing the story	Product dynamic
Functional ownership	Cif (cleaning product)	Power to deal with the toughest dirt every day (Unilever website)	Product performance	High
Emotional ownership	Godiva Chocolate	A timeless beauty, and as pure as Lady Godiva	An emotional story added to the product	Low
Identity credibility	Nike	If you have a body, you are an athlete Innovative, forward-looking and provocative individualist	Product has strong personality traits that are associated with the brand Frequently release innovative products	High
Experience	Disney	A world of magic	One of many touch-points where the consumer can experience the brand	High/low
Cultural credibility	Harley-Davidson	Freedom	Product iconic Added cultural story that explains why it is relevant	Low
Creative credibility	Facebook	Social media platform	People are creating the content/product and the stories	High

Many brands will also have a combination of functional, emotional, self-expressive and cultural intents. Nike could be described as having a 'self-expressive' approach, but looking at the brand more deeply, Nike is using elements from all of the categories. For the Nike brand, the personality of the brand plays an important role. In its category, Nike 'owns' these emotional traits. To use a brand for a self-expressive purpose, the brand will have to earn credibility and ownership of this trait. Nike is about innovation and being forward-thinking. In 1999, the company established Nike ID. Nike ID offers its customers an opportunity to customise their own shoes from templates. Nike was a pioneer in establishing an online system that allowed people to engage in customisation of products. Nike facilitates people's creativity by offering them a platform and a process. This means that the Nike brand can be understood in terms of all of the categories listed above, which proves the importance of defining one's own brand story.

In Table 1.1 the chosen brand strategy gives a clear indication of the role of the product in performing the brand story, and also how active a role any new product could play. When a product takes on a symbolic role, it is important that any product change is carried out with care in order to maintain the brand. With brands where the product has a stronger role in performing the brand story, new products may be crucial in keeping the brand promise alive. When changing the strategy, the role of the product will also have to be clarified.

Key Questions

This chapter has described six different branding strategies:

1.   functional benefit

2.   emotional benefit

3.   self-expressive benefits

4.   experience benefits

5.   credibility in performing myths

6.   providing a platform for customers' creativity.

In all of these the product can play a role, but that role will vary depending on the strategy.

There are many questions that need to be answered in order to integrate the product design process as an activity that builds the brand. What brand story does the product belong to? What is a strong product story? How can the product perform the brand story? How will the market and cultural context affect the product and the brand? All of these elements need to be understood. The following chapters will explore these questions in greater depth.

2

History: Value-centredness in Branding and Design

As a designer you need to dig deep into what this brand is about: Why is it a brand? What is the philosophy behind that? If not, the product will be massively mismatched. When the customers look into it, and they will, they will find out if we are trying to pull the wool over their eyes. A lot of industrial design projects used to be like that. We brought many products in from different suppliers; we tried to make them look the same with colour, graphics and just putting the same brand name on them. I do not think that washes any more. You need to look at what brands are about or else they are not going to sell.

Damian Mycroft, Head of Industrial Design, Hewlett-Packard

This chapter will provide a historic perspective on some of the changes that have happened in both branding and design, and will help us establish an understanding of these two fields. The chapter will also show how these two fields have now become much more similar to each other than they have been in the past.

Towards Brands with a Purpose

In 2009 the advertising giant Leo Burnett launched its new proposal for how to build a brand (Leo Burnett 2009). The company named its approach 'Human Kind', which references the reason given for launching a new approach in branding. According to Leo Burnett, we have been through the 'product era', where it was a question of communicating product features, the 'brand era', where the focus was on the logo, the 'idea era', which was about communicating creativity, and have now entered the 'people era'. In the people era, people are at the centre of our mindset – and brands must have the purpose of making a difference in people's lives.

An academic presentation of these changes is given in the sociologist Douglas Holt's article 'Why do brands cause trouble?' (2002). Holt explores different branding strategies from 1920 onwards, and presents three different paradigms for what a brand is and how companies are building brands: the 'modernist brand paradigm' (1920–60), the 'post-modernist brand paradigm' (1960–2000) and what Holt describes as the 'post-post-modernist brand paradigm' (from 2000 onwards). This section of the book will follow these three paradigms, and will briefly explore the role of the product in building the brand.

IF WE KNOW WHAT TO COMMUNICATE, WE CAN SELL THE PRODUCT

The modernist branding paradigm from the 1920s 'is built on two pillars: abstraction and cultural engineering' (Holt 2002). This paradigm represents a very rational approach to communication. Advertisers had a belief that if a company managed to find out exactly what made people act, consumers could systematically be instructed to value the brand. Tools and methods to tap into consumers' minds were developed in order to understand them. The focus group was one of these tools, still in use, to gain an understanding of what customers want. In this method, groups of potential buyers are invited to discuss certain issues in their lives, and this is fed into developing the marketing campaign.

The 1950s saw a development of different channels that supported companies' need to project a message about their brands. Television started to become part of people's home life, which led to a revolution in the efficiency of spreading such messages. Not only was it possible for the brand owners to market the product in a coherent way, it was also possible for them to enter consumers' households at a time when they were relaxed and perhaps more open to new ideas. Commercials became an important touch-point with the consumer where the company's intended meaning of the brand can be shared. They are strong because of their ability to communicate a complete story of what the brand is about and to use both visuals and sound in communicating this (Holt 2004).

The branding process in the modernist brand paradigm can be characterised as a one-way monologue (see Figure 2.1). In this illustration there is someone who projects the message that becomes the brand, and the customer perceives the message. The message is not necessarily related to the company or the product behind the brand.

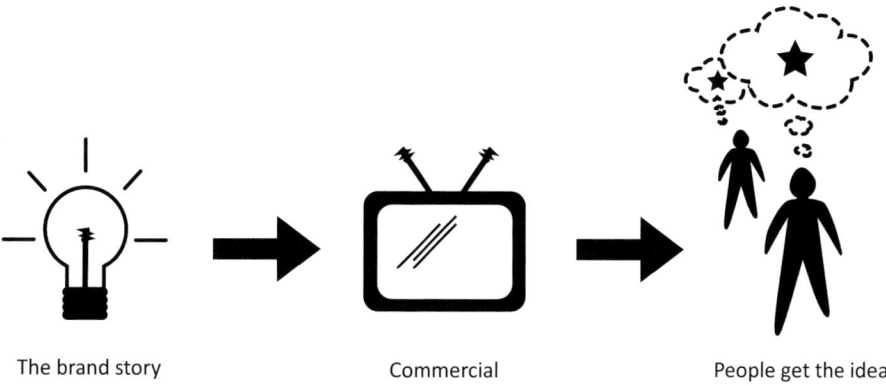

| The brand story | Commercial | People get the idea |

Figure 2.1   Branding as a monologue

The modernist brand strategists' belief in cultural engineering goes hand-in-hand with parallel theories in mass manufacturing. A famous quote from the innovative business leader Henry Ford is: 'Any customer can have a car painted any colour that he wants so long as it is black' (Ford 1923). This points to an attitude in the 1920s that the companies knew best what people wanted. Branding therefore became a value-adding exercise after the product had been developed.

The modernist version of branding represented an abstraction of the message from the product. This approach to design and branding can also be seen today. The research and development department comes up with an idea, the product is manufactured, and at the end of the product development phase the marketing and branding activities 'add value' to the product.

IF OUR COMMUNICATION IS HOLISTIC, WE CAN SELL THE PRODUCT

The second branding paradigm was born in the 1960s, at the time of the emergence of a strong counterculture opposing consumerism (Holt 2002). In this paradigm people were increasingly using branded products to construct their own identity, and companies had to develop brands that were 'authentic'. In developing their brands in this direction, companies started to realise that it was not only a question of marketing a message that consumers would like to hear. This message also had to be 'authentic'. Being 'authentic' in this context means that the message a company projects has to reflect its true activities.

When a company has an 'authentic' message, and this is communicated by means of commercials and other touch-points, it becomes trustworthy for the customer. This authentic message can, for instance, be focused on the company's heritage, production techniques or the founder of the company. The shift to a more 'authentic' branding strategy can be illustrated as in Figure 2.2.

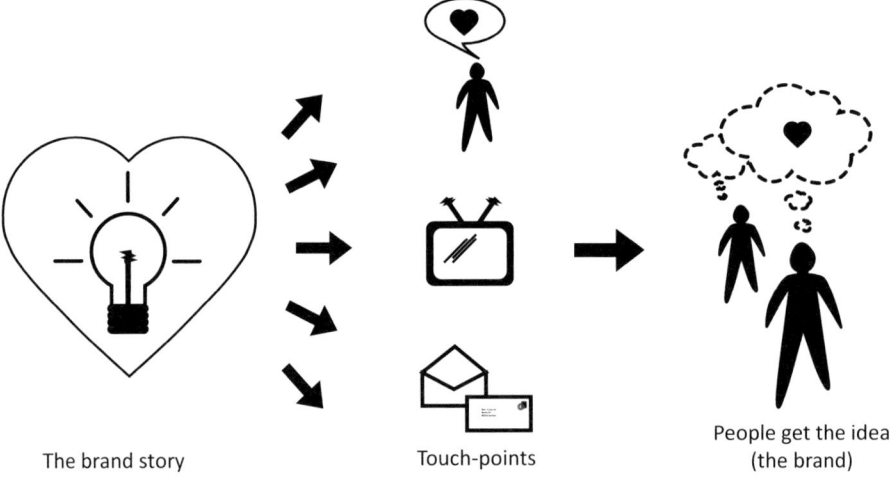

The brand story   Touch-points   People get the idea
                                                                    (the brand)

Figure 2.2   An authentic branding strategy

In the 1960s the British design company Wolff Olins established some of the core principles in how graphic design can play a part in building identity through the development of corporate identity programmes. This introduced a systematic approach to building a brand through coherent use of symbols such as logos. Many companies successfully developed a corporate identity that communicated certain elements that represented the brand in a coherent way.

From developing a coherent corporate identity emphasising graphic design, the focus shifted to developing a coherent identity that includes all activities the company undertakes. In this shift, the brand becomes something more than part of a communication strategy, instead incorporating the whole organisation. The veteran brand strategist David Aaker, a former professor at the University of California, Berkeley, developed the Brand Identity Planning Model for building a strong brand identity (Aaker 1996). This model has had a significant impact on professional brand management.

Aaker's model takes four main perspectives that define the brand identity:

1. the brand as the product

2. the brand as an organisation

3. the brand as person

4. the brand as symbol.

At the core of every successful brand, he says, there is a vision and a mission, and the fact that the brand also has some core values. An important part of the model is the *value proposition* – the functional, emotional and self-expressive benefits a brand is offering to customers (Aaker and Joachimsthaler 2000). The core of the brand identity is communicated coherently in all aspects of the branded organisation.

In Aaker's model, the product is one of four perspectives communicating what the brand is about with its customers. Using Nike as an example, the shoes share the personality traits that also characterise the brand personality. Since the establishment of the brand name Nike and the 'swoosh' logo in 1972 there has been a strong drive to define what the brand is about. In Nike's case, the philosophy has been: 'better technology leads to better performance'. Technology and innovation have been part of the core of the brand, in addition to tapping into telling the story of the strong individuality of top athletes. The athletes Nike has associated with have characteristics like: 'edgy, provocative, aggressive, independent, someone with an attitude – in short, his or her own person'. Nike trainers are often brightly coloured. Asymmetric stitches signify dynamism and play with references to speed or strength. This makes for a dynamic appearance. The logo of the brand, the 'swoosh', is also dynamic. It is asymmetrical and has sharp edges at each end.

In this paradigm, the brand becomes more important than the simple product. One sign of how the brand itself grew in importance is the large logos that dominated clothes design in the 1980s. During parts of this heady decade it was fashionable to wear sweaters with a prominent logo as the decoration. This was not about the sweater in itself, but about the *branded* sweater. The brand was not merely an identifier for trust, it became something that told a story about the status of the wearer. Using brands as a means of self-expression was important.

In the example of the sweater, the product itself becomes less important than the logo it carries. However, the relationship is no longer as straightforward as in this case. Cultural studies of luxury brands and codes within the luxury market show that there is an intricate relationship between the product and the consumer. A product with a large Prada logo is a Prada product. But a Prada product with a small logo is the 'most Prada'. As Professor James B. Twitchell at the University of Florida puts it:

> *The absence of all the usual indications – with the logo hidden, with no design element so distinctive and consistent that it could serve as a signature for the brand – means connoisseurship becomes knowing how to read a secret code.*
>
> *(Twitchell 2001: 93)*

In the post-modernist paradigm there is a shift not only in how the company builds its brand, but also in what the brand means for customers. The understanding is that the story of the brand is not only what the company says it is, but what the consumer says it is. As Figure 2.1 shows, in this paradigm the conversation with the consumer is still characterised as a one-way monologue rather than a dynamic interplay. However, this was about to change.

IF WE HAVE CREDIBILITY, OUR PRODUCTS ARE PART OF THE GAME

Douglas Holt argues that there was another paradigm shift as we entered the twenty-first century. Similar to the situation in the 1960s, a new counter-commercial agenda emerged around the year 2000. One of the most influential contributions was the book *No Logo*, written by Canadian social activist Naomi Klein (2001). This is a critique of both brands and society in general. The book became highly influential in the debate about the development of society (*The Economist* 2001). However, the importance of companies building brands has not diminished as a result of this focus. Quite the opposite has happened. Companies, but also a wider sector of society, have become more focused on brands than ever. There has been a shift in branding that has resulted in an emphasis on how brands play an important role in culture, and brands have thus increased in importance.

The American anthropologist Grant McCracken (2005) and business executives Gene Pressman and Noah Kerner (2007) are some of the other authors who have focused on the importance of culture in brand-building over the last decade. McCracken (2005: 175) emphasises the importance of understanding the cultural context the brands and consumers are part of, and suggests the benefit to companies of introducing systems for managing the meaning the product or service represents. Coming from an anthropological background his main perspective, however, is the important role of meaning provided by marketing for consumers.

> *The consumer has access to many meaning sources beyond the ones provided by marketing. But the ones provided by marketing are vital to the self-invention or self-completion of the individual. [...] Without the meanings made available by the marketing system, the individual is, for some social and cultural reasons, incomplete or at least pallid.*
>
> *(McCracken 2005: 178)*

What characterises this shift in brand-building practice when the cultural dimension is added, is far more complexity in the building of brands. Consumers know that companies build brands for the sake of profit, and they do not have a problem with that as long as the brands have something that makes them credible (Holt 2002). Brands are part of the cultural web. If a brand wants to be associated with a specific culture, it has to have credibility. The importance of credibility and a realisation of the dynamic interaction between people means that brands still need to appear 'authentic'. This dynamic interaction is what Marketing Professor Michael Beverland at the University of Bath builds on in his theory of brand authenticity. Beverland (2009: 5) says that 'brand authenticity is actually derived from an ongoing interaction between the firm, its stakeholders and society'. His understanding resonates with Holt's, in that what makes a brand 'authentic' is not a constructed identity, but an interplay. In this interplay the company projects, and people respond. This could be seen as a cultural dialogue that both validates the brand story and changes it. In this discussion the validity of the authenticity becomes real. The story is not added at the end of the product development process, but is created in dialogue with consumers. Branding becomes a dynamic exchange, with multiple stakeholders, that is centred on values.

Values in this context will relate to an activity that appears meaningful or desirable for people's lives. In a commercial context, the hope is that meaningful desired activity will be rewarded by economic value. This economic value will only be realised because other members of society recognise the activity, or the result of the activity, as something that creates value in their own lives.

The process starts with company initiatives like releasing a product, starting a campaign or a similar activity. Consumers respond to this and interact with the brand story and each other (see Figure 2.3).

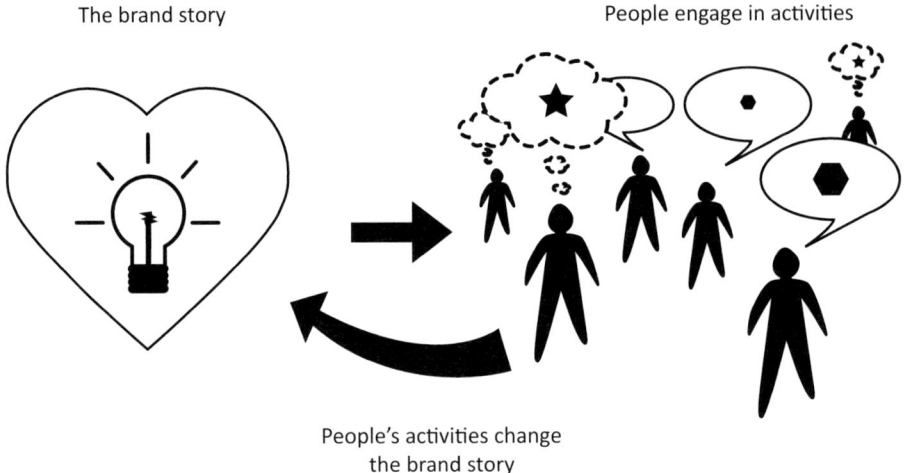

Figure 2.3   **Branding as people's activity**

The brand is strengthened if more people buy into the same story, so that the story is reinforced. If the brand identity is fragmented in the planning phase, the brand will miss an opportunity and be weakened instead of strengthened. To build a brand, a company needs to know how people both inside and outside the company talk about the brand. This may include how designers and marketers communicate the brand, how a journalist explains what the brand is about, or how consumers, either among themselves or in social networks, talk about the brand. In this dynamic exchange of information the product becomes an important activity in building the brand.

The product will play an important role in the dialogue the company has with its consumers. In this dialogue, product design can communicate both the past and the future. The values that people in the company focus on drive the process of making the product, but this is informed by insights from within the industry. The fashion industry has adopted this approach for years, as the enterprise is often centred on the designer. As Karl Lagerfeld told *The New York Times* in 1976:

When people ask me what I do, 'designer' seems inadequate. I tell them that I'm in the fashion business. But that is what happens with ready-to-wear. You become an enterprise.

(Morris 1976)

This approach is also adopted in other companies that are centred on products. In the next section we will look at some of these companies that have actively used strong ideas and values that are important in the design community, and have managed to build strong brands that consumers are interested in engaging with.

How Design Philosophies Have Built Brands

Just like brand theories, design theories can be divided into the categories modernist and post-modernist. In design, modernism started around 1900, and post-modernism in the mid-1970s. Both of these '-isms' represent major philosophies that have played a special role in the design community. There are differences between how American and European designers work. American designers identified the valuable role design plays in a commercial context before their European counterparts did (Julier 2000). This following section is based on a Northern European and Scandinavian understanding of how design philosophies serve as part of the brand. It will show that although designers may not have been involved in brand-building per se, companies that aligned their business strategy with their design philosophies managed to build strong brands.

BRAUN AND THE MODERNIST DESIGN PARADIGM

The modernist paradigm starts around 1900, and is known by different names, such as 'functionalism', 'the modern movement' and 'international style'. In architecture, this style has been characterised as the 'international style' or 'international modernism'. The independent academic Amy Dempsey says that modernism 'is characterised by neat rectilinear forms, flat roofs, open interior spaces, lack of ornamentation and use of new materials and technologies' (Dempsey 2002).

Modernism arose in parallel with the late stages of the Industrial Revolution. The rhetoric that accompanied modernist design was that architecture and products should represent the modern period, which was seen as the

'Machine Age', and that products' form should reflect their functions. The idea was that industrialisation at this point in time was about to bring about a better life for a majority of citizens, and through the use of manufacturing designers and artists could reach a large audience with their designs. It implied a break with former traditions of craftsmanship, and emphasised a style of abstract geometric shapes and clean surfaces in contrast with earlier decorative approaches. The aesthetics of these objects referenced the machines that were used to make them. Design with figurative expression was not necessarily appreciated as 'good design'.

In Europe, the design organisation Deutscher Werkbund ('German Work Federation', founded 1907) and the design school Bauhaus (established 1919) played a major role in defining the principles on which design should be built (Dempsey 2002). The reason for establishing Deutscher Werkbund was to change the direction of German design. Here, the notion of 'good form' was important. The German and Scandinavian design movements were closely linked. In Scandinavia, the Swedish manifesto *Vackrare vardagsvara* ('Beautiful Domestic Items', Paulsson 1919) was largely inspired by the Deutscher Werkbund's motives and visions. This manifesto argued that artists should participate in industry. This would produce 'beautiful objects' that could reach a large audience. However, a moral factor was also included as a reason for engaging in industry: if working-class people could be surrounded by beautiful objects, it was believed that they would also become morally upright citizens.

The modernist agenda had a purpose, and can be seen as being centred – at least nominally – on 'people needs'. However, the modernist movement was a top-down approach. The principles for 'good design' were defined by the design community and often projected to the audience through the use of mass manufacturing. The modernist theory of design gave designers an excuse to design on their own terms while claiming that their forms were the result of objective determinants (Michl 1995). The designers had strong ideas about what was in people's best interests, and these ideas were not developed in relation to users. Users might be brought in to optimise products' functionality and usability, but were not included when it came to defining the aesthetics of the objects.

One of the later companies that can be identified by its strong links to the modernist design language is the German consumer products manufacturer Braun. Dieter Rams was appointed as chief designer for Braun in 1961, a position he held until 1995. During these decades when Rams was chief designer, the company developed a distinct product language based on the functionalist design language and principles.

In the exhibition catalogue for the exhibition *Less and More: The Design Ethos of Dieter Rams*, Rams's functionalist design principles are listed as 'ten commandments'.

Good design:

1.   is innovative

2.   makes a product useful

3.   is aesthetic

4.   helps to understand a product

5.   is unobtrusive

6.   is honest

7.   is durable

8.   is consequent to the last detail

9.   is concerned with the environment

10. is as minimal as possible.

These 'ten commandments' are general rules, and have had a strong influence in defining industrial design. These principles can be interpreted in many ways, but looking at Rams's products, they have given a clear direction for the design process. His principles that good design should be as minimal as possible and unobtrusive indicate a functionalist concept of form, where the form of the product should follow the idea of the function: in other words, the designer should rely on technical forms for aesthetic effects. Decoration for mere aesthetic purposes is not seen as good design. The principle of honesty is a subjective one – but gives indications for the materials to be used, and in relation to the other principles also indicates development of a functionalist form.

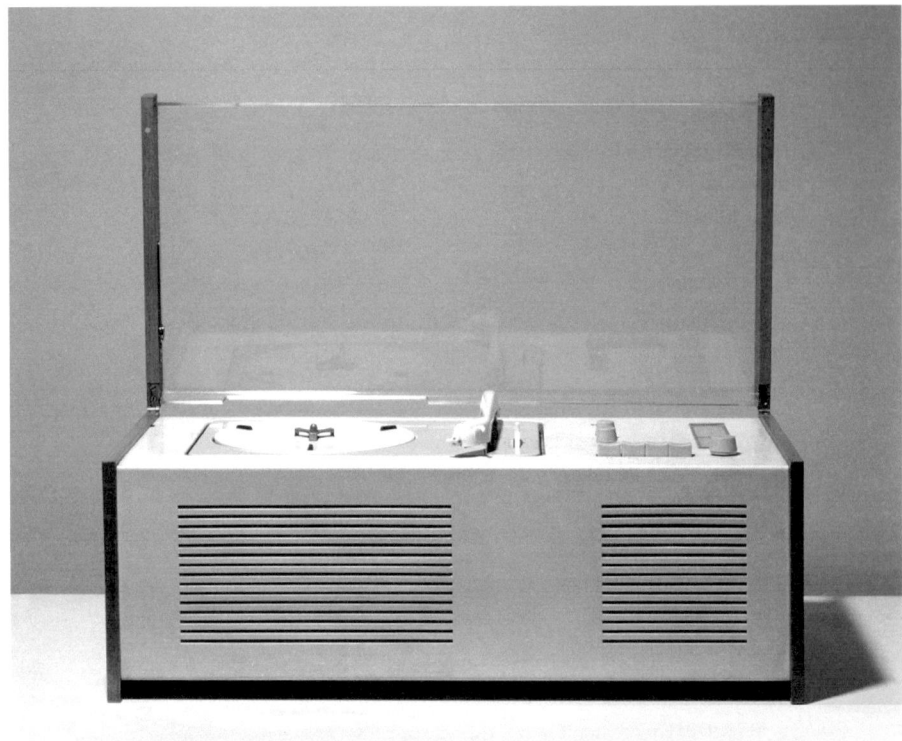

Figure 2.4   Braun SK4 record player by Dieter Rams and Hans Gugelot (1956)
Copyright Braun GmbH.

How passionate Dieter Rams was about his designs was described by his colleague Rudolf Schönwandt. According to Schönwandt, Rams considered input from the marketing department almost a personal affront. Schönwandt described how it was easy to see what got on Rams's nerves: 'extreme visual stimuli, … obstructive colours and shapes that aggressively demand our attention' (Burkhardt and Franksen 1980: 21).

In Scandinavia, the 1950s was an important period in establishing what has later become recognised as 'Scandinavian design' (Skjerven 2003). Some of the values that endure in the Scandinavian design community to this day emerged during that period: values like durability, honesty and closeness to nature. These values remain subjective, and honesty in particular has become an ambiguous concept. Scandinavian design is perceived as one concept;

however, there were distinctions between the designers in this period. While Swedish designers were influenced by Bauhaus, as described above, Danish design placed a greater emphasis on craftsmanship. The role of this in Danish design ran counter to many other modernists' emphasis on industrialising design projects or focusing on functionalist objects.

One brand that had a particular influence in defining the idea of Scandinavian design was the Danish silverware company Georg Jensen, which has a prominent heritage in producing high-quality silverware. The Danish designer Henning Koppel worked for Georg Jensen, and designed some of the objects that remain unique products in the company's portfolio. Koppel's background was not in design or jewellery, but in fine art. This has later been seen as a reason for the sculptural quality in his work (see Figure 2.5). Although Koppel was following a design language that was 'modern' and a break from ornamented products, his work emphasises form, not functionality or ease of production. However, his work was of significant importance in bringing the Georg Jensen company into what was later perceived as 'Scandinavian design'.

Georg Jensen has developed a product portfolio with distinct products that somehow belong to the same family but do not share the same explicit references, nor are they all designed by the same designers. However, the products share some of the same characteristics. They are all part of the Danish design tradition, and one aspect of this is a high degree of finish. The distinct profile of Georg Jensen is something the company has been able to preserve despite employing various designers with strong profiles of their own, including Arne Jacobsen, Sigvard Bernadotte, Nanna Ditzel, Henning Koppel and Vivianna T. Bülow-Hübe. In Georg Jensen, the values the company developed in the course of adapting its portfolio to a modern Scandinavian design language are still part of what the brand is about, and are visualised throughout all the company's marketing materials.

Both Braun and Georg Jensen developed strong brands by letting key designers develop their philosophy in the products: Braun by having a strong design leader internally, Georg Jensen by calling on multiple designers with shared values. The systematic use of design to communicate a message has strong resemblance to the strategy for building a brand. The aesthetics of the products, followed up by the promotion of these products, builds a visually coherent impression of the brand. Both of these brands have been pioneers in bringing designers into their organisations and developing a designer-led approach to brand strategy.

Figure 2.5   Pitcher by Henning Koppel for Georg Jensen (1952, 2008)
Copyright Georg Jensen A/S.

ALESSI IN THE POST-MODERNIST DESIGN PARADIGM

In the 1970s and 1980s there was an 'anti-design' movement within the design community that criticised the principles of modernism (Dempsey 2002). This movement questioned the notion that design was only about functionality, and emphasised the social and cultural roles of products. The Italian designers who formed the Memphis Group, led by Ettore Sottsaas, were important in changing the concept of design (Design Museum 2011). At the Design Fair in Milan in 1981 they exhibited pieces that strayed far from the functionalist approach to design, and experimented with a wide spectrum of colours, exuberant shapes and low-end materials. This group was important in what has later been understood as the post-modernist approach to design.

One of the companies that played a part in changing the discourse about design is the Italian kitchenware manufacturer Alessi. Its stated mission is that the features of its products are not there purely for functional reasons, but exist so that consumers can change their perceptions of their homes and society (Alessi 2012). In Alberto Alessi's own words:

> *transforming the gadget function ascribed to objects by the consumer's society into a transitional opportunity, namely into an opportunity for consumers to improve their perception of the world .... In the future most of our products will continue to be marked by a high degree of innovation and experimentation, as we believe this is the way to develop our ability to set trends, to promote our fame and to create a culture medium aimed at developing those projects we like to call Super & popular.*
>
> *(Alessi 2012)*

One of these Alessi products is the Juice Salif lemon squeezer designed by French designer Philippe Starck. It has been criticised for having poor functionality, but the real functionality of this object is open to debate. The Juice Salif can function as a decorative object or even a conversation piece (Snelders and Lloyd 2003). It conveys the idea of a spaceship, the legs of a robot or an alien figure.

Alessi got involved in the design discourse by inviting ten leading designers to design a kettle (Verganti 2010). This first project was not for a commercial purpose, but in order to design a prototype for museums. Along with the design, Alessi also published a book about the design of the kettle. Ten prototypes where sold to various museums. Later, Alessi approached the

American architect and product designer Michael Graves with a new brief, in which he was asked to design a product that was suitable for the mass market. The object he designed was the Bird Kettle. The functional aspect of the kettle is to boil water, but the little bird whistling when the water boils adds a new meaning to the kettle. This type of innovation changes what a kettle is about – a shift from focusing on functionality to focusing on emotions.

Because of Alessi's efforts in disrupting the discourse of what is acceptable as 'good design', the product was also well received.

Alessi's strategy is different from Braun's. Braun had one chief designer with strong design principles. Aligned with the essence of post-modernism, not concerned with having one voice, Alessi has developed a web of interpreters (Verganti 2010) who are making tangible objects with cultural meaning. By working with a wide range of interpreters, Alessi manages to maintain a dynamic and up-to-date brand.

Alessi shares the systematic approach of Braun and Georg Jensen, but applies it differently, particularly compared to Braun. At the core of Alessi's activity there is a desire to design products that reflect changes in culture and give a new dimension to the products. The result is playful, and often results in products with a figurative expression. This playfulness, along with the story of the designer, is communicated through all the company's marketing channels. The focus of Alessi's communications is on the products and the designers behind these products. The idea of the product has been reinterpreted in commercials. Alessi has also established close links with museums, and its products can be seen in a wide range of such institutions all over the world, including the San Francisco Museum of Modern Art (SFMOMA) and the Museum of Modern Art (MoMA) in New York. This adds a cultural dimension to its products and bolsters the company's strong position.

Post-modernism in design represented a break with the modernist approaches. Being a pioneer in building on a post-modern design philosophy has granted Alessi a strong brand position. At the same time, many people would not see a great difference between a chair designed in the modernist era and one designed in the post-modernist era. It has been questioned whether post-modernism is a 'paradigm shift' or whether it follows some of the ideas of modernism (Dempsey 2002). The emphasis on communication and telling a story is strong in post-modernism, and also integral to brand-building. Post-modernism questioned the idea that there was one right answer to 'good design'.

In this period it was emphasised that designers should develop products that communicated with customers. How much the user was involved in the post-modernist era can also be questioned. In the example of Alessi, it perhaps didn't directly involve the user to a great degree. All the same, another direction in design began to be developed: participatory design, or user-centred design. In this approach designers listened to users and even designed products in collaboration with them. What the post-modernist design era has opened up is multiple directions the design profession can take next.

Modernism and post-modernism represent two different aesthetic approaches to visual idioms which have been immensely successful for some brands and products. However, in a pluralistic world one style will not fit all. This is not always recognised in the design community. In Northern Europe the principles of modernism still play a dominant part in design curricula (Michl 2007). This is also evident by looking around us and seeing which products are perceived as 'design products'. These have a strong resemblance to the abstract modernist design language.

DESIGN TODAY

Lately there has been a new shift in the design community which has parallels with the contemporary shift in branding. There is a blurring between disciplines. Institutions that used to hold power, such as the organised mass media, are seeing their positions eroded.

The design researcher Anna Valtonen has studied how the role of Finnish designers evolved over six decades. Today some designers take on the role of a pushing innovation and defining the company's vision (Valtonen 2005; Valtonen 2007). Looking at present leading companies, design is integrated as an important asset for the company, and is driven by a strong vision. One of the leading companies in this respect is Apple. Apple is an iconic company with a strong position among designers, and has also gained a very strong financial position (AppleInsider 2010). The powerful visionary leader Steve Jobs represented the Apple brand for many years. By introducing Jonathan ('Jony') Ive as a chief designer who reported directly to the head of the company, Jobs built a strong sense of what 'good design' might be in the 2000s. At Apple, not only did Ive report to Jobs, but both were engaged in the process of making even minor decisions about what the product should look like. The BBC documentary *Steve Jobs: Billion Dollar Hippy* interviewed Apple's former Senior Vice President of Software Engineering, Avie Tevanian. He described the level of detail that Jobs was engaged in: 'Choosing white for the iPod wasn't just a Jony decision. It was a Jony and Steve decision' (Quinn 2011). However, the design of Apple's products has references to Braun and to the modernistic paradigm. Steve Jobs's biographer, Walter Isaacson, describes how Jobs had already publicly embraced the Bauhaus approach at a design conference in 1983, and also 'predicted the passing of the Sony style in favor of the Bauhaus simplicity' (Isaacson 2011: 126). Isaacson goes on to describe how Jobs talked about doing for computers what Braun had done for consumer electronics. Apple's design mantra would live up to the motto in its first brochure: 'Simplicity is the ultimate sophistication' (Isaacson 2011: 127).

Apple design concerns more than the aesthetics of its products. In Apple, the whole user journey is carefully thought through and designed. The iPod music player is an example of a product where aesthetics are important, but so are the user interface, the technology and how it connects to an integrated service that makes it easy to access music desired by the user. The creative human being is at the core of what the products are about. This is also a story that people would like to interact and engage with. It is the holistic picture that Apple designers are part of, and why these products represent a design philosophy that is relevant today.

A more controversial example that represents what design is today is the company Threadless. This represents the shift from the designer or the design team as the sole creative agent to involving everyone in the organisation, as well as customers, as part of the design process that leads to the final product. Novel solutions like rapid prototyping and various software programs make it possible for everyone to become 'a designer'. In this scenario the professional designers become facilitators for other people's creativity. Threadless was set up with the goal of inviting a range of consumers to design their own t-shirts. People voted for their favourite, and the one that gained most votes would be produced and offered for sale. Threadless is a community more than it is a business. The making and promotion of the product is integrated into the offering. Threadless embraces people's own creativity, and has managed to create a process and a platform people can engage with. This approach also goes along with strong design principles in the design community, considering the role of the designer as being a facilitator for other people's creativity, or even seeing every human being as a designer. This design philosophy contrasts strongly with Apple's. Apple has a far more distant and closed relationship with its customers, while Threadless integrates them. However, both are successful businesses with a strong story to tell.

There are indications that the role of design going forward will be far more pluralistic than the one represented in modernism and post-modernism. Both modernist and post-modernist design approaches are factors in building strong brands when used in the right context. Some of the principles defined in these paradigms remain important today. Therefore, shifts are not breaks, but stages in evolution. The rise of the creative class (Florida 2002), where more of the population have creative jobs and take part in the act of creating, has also led to a greater acceptance that 'good design' is what is appropriate for the context, and not something defined by 'timeless' tenets.

Engaging Based on Values

This chapter has taken a historical view of design and branding. There have been parallel changes in these two fields, and they can be understood through three main paradigms. An interesting aspect of the recent move away from post-modernism is that branding and design have grown closer to each other. In the design community, some ideas have had a virtual monopoly in defining what good design is and what the right approach is. The brands addressed in this chapter – Braun, Georg Jensen, Alessi and Apple – can all be seen as 'design-led'. They are driven by values that are embraced by a majority of the design community. Their design work is led by a desire to create products that are meaningful. They have managed to communicate what they are about, not only in promotional material, but through holistic thinking in their approaches. This is communicated in every detail of what the product is about. With closer examination of these products, every little detail, including the finish and packaging, is considered thoughtfully. This is important in building a brand – but it is also one of the principles defined by Dieter Rams that remains very influential among designers.

Brands like Braun, Georg Jensen and Alessi have managed to establish themselves with close links to a particular period, but they continue to be relevant in the new paradigm because they bring something meaningful and purposeful to people. Timing was essential when they established themselves as iconic brands. Their interaction with designers and the devotion to their mission are what will make them relevant in the future. Their constant interpretation of shifts in society through the prism of their own values is a key to their success. All of these brands were pioneers in what they did, and created a unique product portfolio. In order to manage that, they have all required strong leadership.

What has made these success stories, and what is common to them all, is that products have been an important part of communicating the story of the brand. Looking around, there are many strong brands with strong stories. Some of these will be described in Chapter 3.

3

Products: Performing Brand Stories

Brands are realising the importance of the product experience beyond purchase. The LG Prada phone is an example of what can go wrong if they don't. Both the branding and the product design people did a great job. It attracted lots of people. It was a beautiful object and launch advertising and in-store displays were stunning. The phone came in beautiful packaging, almost like a watch. But the usage part of the experience was a disaster, the interface was barely usable and lots of consumers returned their phones.

Not only has it become harder for brands to reach consumers through marketing communications, but people now share their experiences of products online. Other consumers trust what they are hearing online more than what they see on the billboard. This marketing emphasis on fostering positive word-of-mouth recommendations brings brands back to the quality of the core product experience.

Kevin McCullagh, Plan Strategic

This chapter will investigate the relationship between the product and the brand in greater detail. It will present a wide range of different brand stories where the product is an integrated and important part of performing the brand promise.

The Product as Part of Building the Brand

THE IMPORTANCE OF UNDERSTANDING THE JOURNEY

The product is seen as one of many touch-points that can allow the consumer to engage with the brand. Companies have a wide range of touch-points, each playing its part in building the identity of the brand. The product helps to build the brand not only in terms of what consumers think when they see the product, but by becoming part of their interaction with the brand through a journey. In this interaction, the use of the product will play an important role in what becomes the consumers' understanding of what the brand is about. This can be exemplified by buying chewing gum.

A commercial that tells a persuasive story about long-lasting taste, an exceptional flavour and a cool pack elevates the chewing gum experience to being about more than just the gum itself. This idea of the brand is important in encouraging customers to select one pack of chewing gum rather than another. However, when opening the pack and choosing one of the pieces, the gum has to deliver what the brand promises. In experiencing the product, consumers make up their own interpretations of what the brand is about. If the chewing gum loses flavour quickly or the texture is like rubber, it may be enough to put people off buying it next time. Their own experiences with the product become what consumers think about the brand. In this way, the product becomes an important part of delivering the promise of the brand. If consumers are really disappointed, they may even tell their many Facebook friends about their bad experiences.

The process from seeing the advertising to buying the product to using it then disposing of it is called the *customer experience journey*. In brand-building, how the consumer experiences the brand through multiple touch-points along this journey is of critical importance. The interactions at the different stages will all be part of how holistic an experience the user gains from the product (see Figure 3.1). In all of these different stages the company will have opportunities to build an awareness of the brand. Each of these touch-points should be planned together to achieve a holistic understanding of how the company is building the brand.

The customer experience journey can be divided into four stages:

1. pre-sale

2. sale

3. use

4. disposal or reuse.

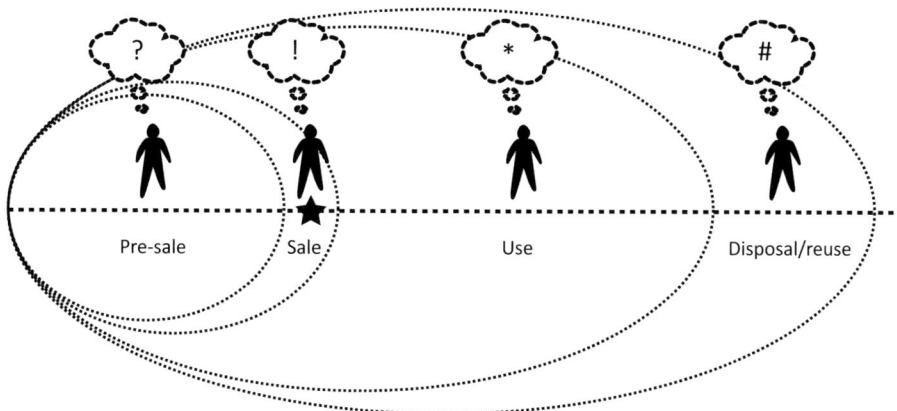

Figure 3.1   The customer experience journey

In the pre-sale stage, advertisements will play an important role in building brand awareness and introducing new products to the consumer. It is important to analyse the pre-sale stage, and the type of product will suggest particular approaches to the journey.

During the different stages, the product will play different roles in building the brand. The buyer's first impression of the product in the shop is important in building the idea of the brand. Physical appearance will play an important role in the shop, both in terms of what the product looks like and also what it feels like when the customer touches it. When the customer sees the product in a shop, it immediately communicates something. The colours, the silhouette, the form – all of the elements as a whole are important in giving the first impression. This first meeting between a potential buyer and the product is an important consideration for design teams. Questions concerning how the product performs in this context therefore arise:

- Does the form evoke the customer's curiosity?

- Does the silhouette make it stand out?

- Does the colour signal the type of product?

- Is the form appealing?

Other elements of the product may become more important when the customer starts to use the product. At this stage, it is important to deliver the promise of the brand. The second time a consumer buys a product, previous experiences with it become an important part of the brand (see Figure 3.2). This is why the branding process is a continuous journey, where the customer's own experiences become an important, integrated factor in building what becomes the brand.

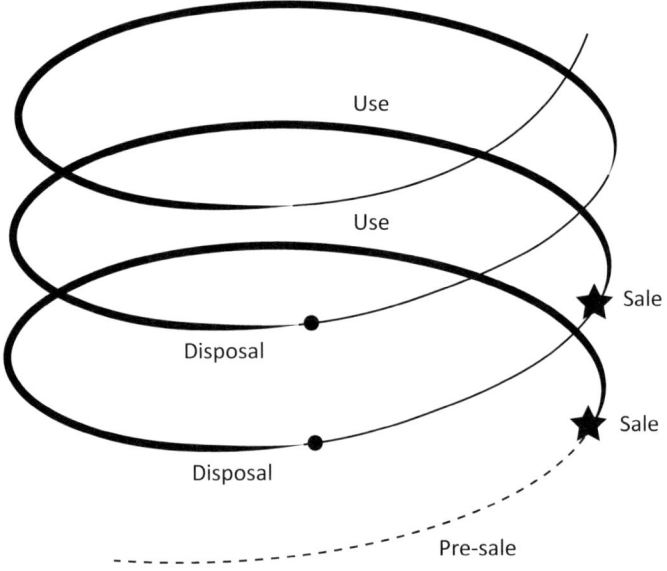

Figure 3.2   The customer experience journey as part of building the brand

This chapter started with a quote from Kevin McCullagh, founder of the British product strategy company Plan Strategic. He shared his perspective on the importance of the product experience. He also emphasised the importance of understanding the category the product is part of in order to understand the experience:

> *If you are talking about the FMCG category, packaging is very important. That is a tangible product. In perfume it is all about the packaging. If you are talking about mobile phones or technology, packaging is less important. You do not tend to see it before the purchase.*
>
> *(McCullagh 2012)*

To make sense of products, they are often described by such categories. Within each category there is what can be called a taxonomy – a hierarchical classification. A convertible car is part of the categories 'convertibles', 'cars' and 'transport'. In design, these categories dictate the product's form. The form is shaped by customer expectations of colours, sizes and similar aspects that define what a product looks like in a particular category. It is also affected by physical requirements created by the environment that the product is part of, such as size of shelves, transport crates or other limitations within the category. By breaking the category rules, new brands can be established. This was the case with the Stokke Xplory pushchair, which will be analysed in Chapter 7. When this pushchair was launched, it succeeded in breaking the established category rules, and it successfully established a new category of high pushchairs.

Computers and other electronic consumer goods were described by McCullagh as 'considered purchases'. The more complex products are, the more likely it is that consumers will have done some research before making up their minds about buying them. In the case of considered purchases, consumers will already have some ideas when entering a shop, but the first impression of the product in the shop may change these preconceptions. McCullagh continues:

> So when people come in they have a rough idea in mind, but there are products and brands in that store that attracts them more than others. People may think; 'the product should be under £600, I am going for these brands and I would like to have a hard disk drive of a certain size.' What happens when they are in the store is that people suddenly get distracted, there are products they do not expect to see. Emotions take over. 'Oh, that was attractive, that was great, this is me.' The physical product becomes important in the sale situation.
>
> (McCullagh 2012)

Understanding the category the product is part of and how people behave in this category is of key importance in developing products that become relevant.

EXPERIENCING WITH ALL THE SENSES

An important part of consumers' experience with the product will not only consist of what the product looks like, but also how the product can be explored with all the senses. The Danish brand theorist Martin Lindstrøm suggests that if brands manage to build ownership of cues that stimulate all the senses, their consumers' loyalty will increase (Lindstrom 2005). Brands should therefore adopt a holistic approach to the senses in brand-building when appropriate. This could be communicated through the sound, the feel, the taste and the smell of the brand. In this, the product can play an important role. A chocolate will have a certain look, a feel when touching the packaging and when feeling the texture of the chocolate in the mouth, a smell when opening the package, a sound when opening the package, and a taste when chewing the chocolate. Developing coherence in how the product is experienced through all the senses will be an important part of building the brand.

The American fruit drink Snapple is one product that has managed to build ownership of a certain look, feel and taste. In this, the design of the bottle plays an important role in building a brand that is easily recognised. The bottle has a characteristic silhouette. The texture of the bottle is also unique and evokes a specific feeling. The Snapple bottle cap is different from other bottles, being oversized. This not only gives the bottle a characteristic look, but also leads to a unique drinking experience. When opening this bottle there is a distinctive popping sound, which also serves as a quality guarantee, indicating that the bottle has not been opened earlier or tampered with.

The product's taste and smell will normally, with few exceptions, fall outside the scope of the designer, as this will require different expertise. Consumers grow accustomed to a certain taste, and the brand is built by coherently delivering the same taste. This is one of the key challenges the food industry has faced since the start of branding. In the fashion industry, the legendary French fashion designer Coco Chanel earned her fortune from the perfume Chanel No. 5, which was released before Christmas 1921. The fragrance of this perfume gives strong references to the brand.

The tactile experience of the product is of key importance in product design. The use of cheap plastic materials will communicate that it is a cheap product. The heaviness of a new camera, the coldness of the surface of a steel fridge, the hard edge on a remote control, the smooth surface of a computer, the edgy transition from one surface to another, how heavy the product is to lift and how the form is designed to be used – all of these are part of the consumer's experience of the product, and important in defining the perception of the brand. During the design process the design team will need to consider carefully how these aspects contribute to building the brand.

OPPORTUNITIES THAT ARISE FROM UNDERSTANDING THE STAGES

Through the customer experience journey customers learn what a product is about. In the pre-sale stage the company tells customers what they can expect. This aspect is important in building an overall impression of the product. In the store the product itself plays an important role in telling the story to the consumer. When the product is bought and brought into use, the user will gain new experiences of what the product is about. In order to understand how the product is part of building the concept of the brand, it is important to analyse and gain a good understanding of all these stages. Companies will find that they do not have control over all aspects of the different stages.

Taking a holistic perspective on the role the product will play at different stages in the customer experience journey also opens up new opportunities. Sometimes powerful brand stories can be created by focusing on one of the stages. The Coca-Cola Company is a powerful market actor that releases many new brands globally. Innovation has therefore proven to be a key tool for Coca-Cola in creating new market opportunities by introducing brands that are differentiated from others in the market.

The bottled water brand market is highly competitive. Coca-Cola created the I Lohas brand when it wanted to launch water on the Japanese market (I Lohas 2011). For this it created a new bottle that focused on a particular user scenario. The bottle was fully recyclable, and partially made from plant-based materials that reduce dependence on non-renewable petroleum resources (The Coca-Cola Company 2010). The form of the bottle communicates the ease and fun of recycling by letting the consumer twist the bottle after use. By innovating the packaging and following up its message in the promotion of the new brand, Coca-Cola managed to introduce a new brand in a saturated market (Keferl 2009). Coca-Cola's successful introduction of a new type of bottle shows how important it is to understand customer experiences at all stages in the consumer journey. The I Lohas bottle also shows that user scenarios can be a powerful customer communication tool when the idea of the product is aligned with these scenarios. When these references are aligned with what the brand is about, the brand becomes a trustworthy one that users tend to remember.

Strong Brand and Product Stories

BRAND STORIES WITH SUBSTANCE

In the book *Brand Leadership* the Berkeley, California-based branding professor David Aaker warns brand managers about the product-attribute fixation trap (Aaker 1996: 72). This is the most common hindrance to building a strong brand. The trap consists of fixating on a simple set of attributes associated with the product and believing that these attributes are the only relevant bases for consumers to make decisions. Such attributes may be a quality or a feature associated with the product, one example being Volvo's focus on safety in its cars. According to Aaker, focusing on attributes is a trap because attributes can be generic and fail to differentiate the brand, may be easy to copy, assume a rational consumer, limit the ability to extend the brand to other categories, and reduce the company's ability to expand the brand to other markets.

The brand story is important in building the brand. Focusing solely on the product is simply not sustainable for a business. A business is so much more than the individual products it manufactures. However, it is important to keep in mind that the product is a vital part of the brand story-telling. By designing products where the qualities and characteristics are aligned with the brand story, a company can create a story with substance. This is where product design becomes important in branding. In designing the product, the design team can define a strong product story. Designing a product that is aligned with what the brand should represent for the consumer will create recognition, emotional attachment and relevance for the consumer. In strong brands the product often plays a distinct role in communicating what the brand is about.

Both stories about the product and stories in the wider context have been used successfully in building brands (Thjømøe 2003). It is important that the company is not too ambitious in trying to communicate too many stories with one brand, or that there is a lack of coherence in how the product represents the brand story. Too many stories associated with a brand or inconsistency will over time lead to fragmentation, and the brand will appear less strong. The company and the design team will have to commit to a particular brand story. By focusing on a core message the team would like the product to represent, this idea can drive the design process. This does not mean having a single message that is repeatedly applied to different products, but being passionate about an idea and letting this idea drive the development process.

Figure 3.3 presents an overview of various product stories that can be used to tell a story about a brand, although this model does not cover everything.

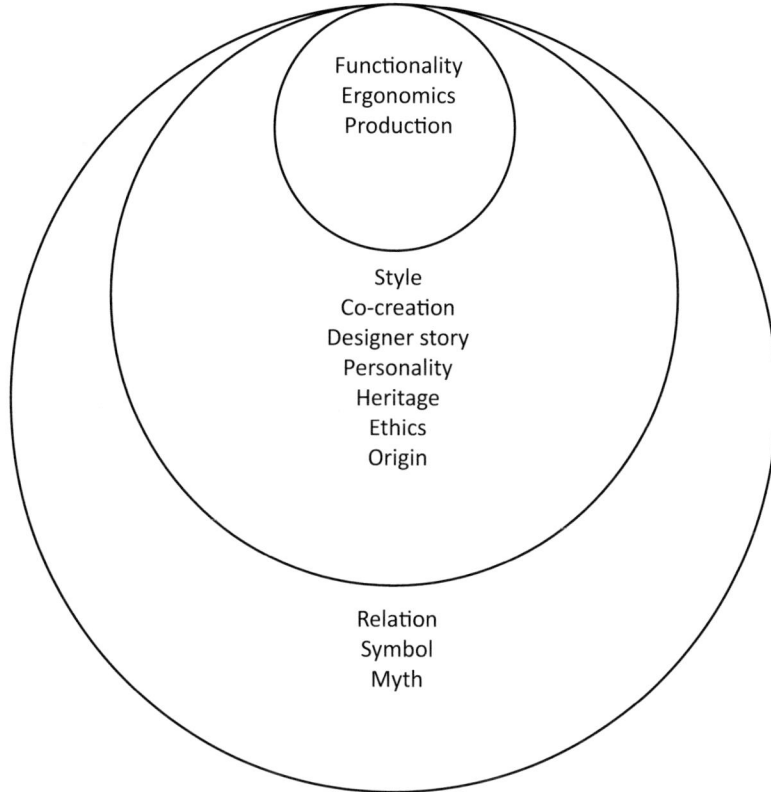

Functionality
Ergonomics
Production

Style
Co-creation
Designer story
Personality
Heritage
Ethics
Origin

Relation
Symbol
Myth

Figure 3.3   Three levels where the product can tell stories about a brand

The model consists of three layers. These three layers represent three levels of abstraction from the tangible product. The first layer contains stories about the product. The second shows these stories in a wider context: the product tells a story through its form, but this story is not necessarily related to the product itself. The third layer contains cultural stories. In these stories the product has a more symbolic role. We will explore a few examples of brands that have succeeded in developing products with a strong story that builds the brand. The brand story may be related directly to the story the product is communicating, but the story the individual product is telling is often part of the bigger picture that the company is trying to communicate about itself.

STORIES ABOUT THE PRODUCT

Products can play an important part in the brand stories (Thjømøe 2003). Stories that are directly related to the product include introducing new functionality, or how easy or comfortable it is to use the product.

Functionality

A few brands have managed to establish themselves based on bringing a completely new product to the market. An example of this is the Post-it notepads. Post-it has become an important sub-brand for the American conglomerate 3M. 3M has very strong product associations and has created the archetype of the product category. Products from 3M have a technical look and feel which is associated with communicating the intended function of each product. This fits well with placing innovation at the core of the Post-it brand, and Post-it's parent company, 3M, prides itself on being one of the world's most innovative companies.

Breaking the norm in the product category or introducing a new product category can be the prime motor in the development of products. When bringing a new product to market, it is of key importance to communicate the user scenario and why the user scenario is relevant. The *raison d'être* of the product itself needs to be explained when promoting the product. Sometimes a product becomes so closely linked with a brand that the brand consists of only a single product line.

Ergonomics

The Finnish brand Fiskars is built on company traditions dating back to 1649, but particularly focused on the ergonomic scissors with an orange handle, designed in 1967. The handles of Fiskars' scissors communicate how to use them by referencing the palm of the hand. The look and feel of the ergonomic scissors from Fiskars constantly reminds the customer that Fiskars is a brand for ergonomic products.

The Swedish manufacturer of professional tools Bacho has also built its reputation on ergonomic tools. The handles of Bacho screwdrivers communicate efficiency in use through their contrasting colours and characteristic grips.

What is important about the approach taken by Fiskars and Bacho is not only the fact that the products communicate that they are ergonomic, but the user experience, where these products fulfil the brand promise. The products must also continue to be ergonomic products.

Production

Each product that is manufactured can also reveal signs of how it was produced. A seam along the body of a plastic product tells a story about where the mould was divided. An indentation underneath a bottle tells a story about where the plastic was injected into the mould. In many products these signs are hidden or made as invisible as possible to the user, but in others these signs of production can make the product look more authentic. Wine from the French brand J.P. Chenet is mass-produced but comes in an asymmetrically shaped bottle similar to those in use when they were produced by hand. However, J.P. Chenet was established in 1984.[1] In contrast with the former hand-made era, every single one of these modern bottles now has the same shape. This has become part of the brand references.

STORIES IN THE WIDER CONTEXT OF THE PRODUCT

The story the brand communicates can also derive from the wider context of the product (Thjømøe 2003). This could be a narrative that is related to the product, but does not have to be communicated through the product itself. However, in strong brands products often communicate a story.

Style

Technology changes, materials change, and user scenarios change. This can present both advantages and disadvantages in building brands. The shape and the form of the product give an indication of when the product was designed. In design there have been periods where certain styles have had stronger influence than others. This is particularly true for electronic devices, as there is a constant drive to adopt new technologies.

The Danish upmarket electronics producer Bang & Olufsen (B&O) produced the BeoSystem 2500 (see Figure 3.4). With its clean and geometric form, the design was seen as 'timeless'. However, the form still gives an indication that it was designed in 1991. More important is the radical change of technology in the music industry, with LP records being overtaken by cassettes, and compact

---

1   My appreciation to Professor Jan Michl for providing me with this example.

discs (CDs) being overtaken by non-physical music media such as mp3 files. B&O is an example of a company that has succeeded in reinventing a style from one epoch to create a distinct product design. The BeoSound 8 (see Figure 3.5) was launched in 2010, almost two decades after the BeoSystem 2500. Nevertheless, both products were designed by David Lewis and they show strong brand affiliation. The products belong to the same brand, although the functionality and the technology they support are different. By keeping up-to-date with changing technology while remaining true to the geometric form language, B&O has created a brand with products that have strong characteristics. The clean surfaces and geometric shapes of B&O audiovisual equipment are aligned with the company's devotion to style and its passion for music.

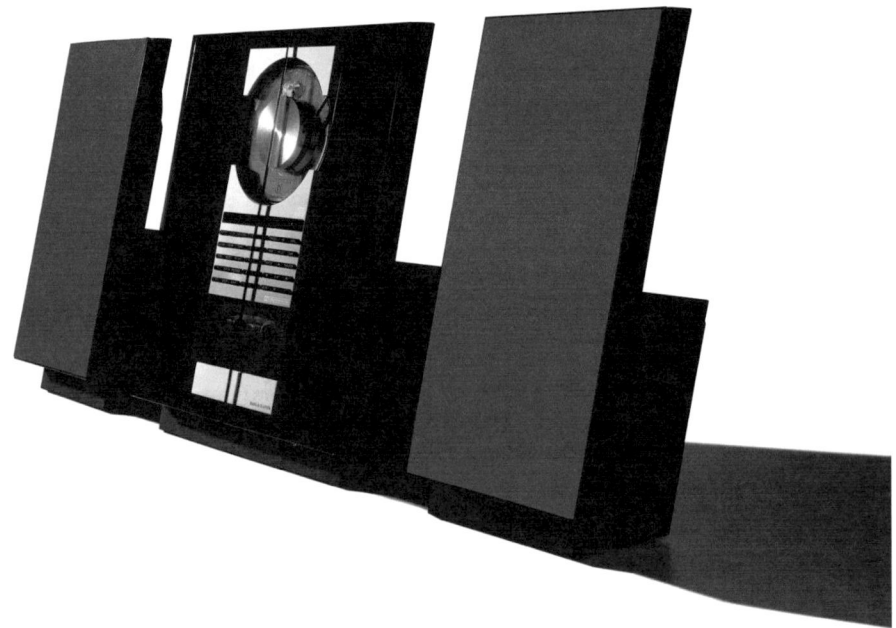

**Figure 3.4   BeoSystem 2500, designed by David Lewis for Bang &
              Olufsen (1991)**

Copyright: Bang & Olufsen A/S.

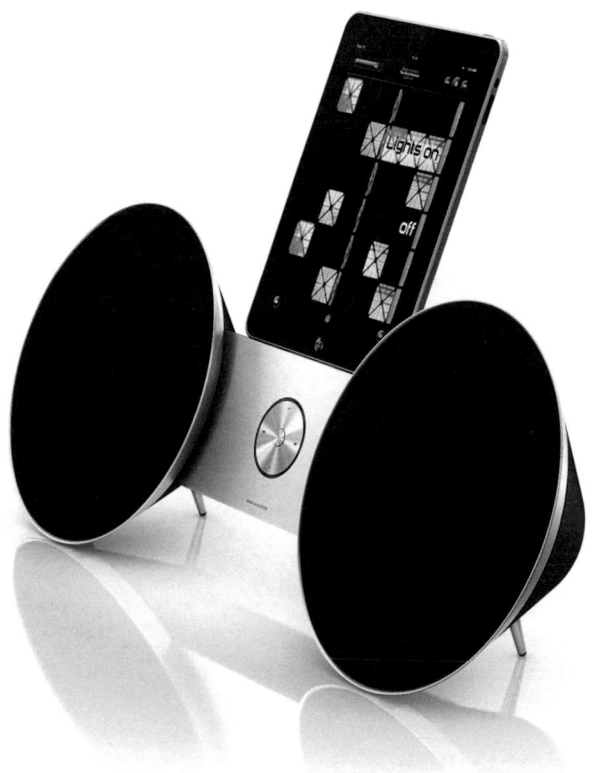

Figure 3.5 BeoSound 8, designed by David Lewis for Bang & Olufsen (2010)
Copyright: Bang & Olufsen A/S.

Heritage

The German camera manufacturer Ernst Leitz introduced the Leica M3 model in 1954 (see Figure 3.6). Leica has become a strong brand, based in particular on the company's devotion to this camera model and later iterations of the same design. The photography industry has experienced dramatic technological changes, particularly with the transition from photographic film based on chemicals to digital technology. Despite this technological revolution, Leica chose to launch a new 'M' model for digital photography in 2006. The Leica M8 (see Figure 3.7) has strong references to the M3 camera introduced in 1954. This strategy of preserving the heritage of the legendary design provides contemporary Leica cameras with an authentic feel, and communicates that the camera has a strong heritage that the manufacturer is proud of. The Leica

M-series cameras are also very close to what is considered the typical form (type form) for a camera. The preservation of this form, with products adapted to the latest technological developments, tells a story about a brand that is proud of having delivered high-quality cameras for decades and being part of defining what a 'camera' is.

Figure 3.6   Leica M3 (1954)
Copyright Leica Camera AG.

Figure 3.7   Leica M8 (2006)
Copyright Leica Camera AG.

Co-creation

The trend of inviting people to co-create with the company, or seeing the company as a platform for creation, has developed new, powerful brand stories about the people who make the products and designing as a collective process. The t-shirt brand Threadless, mentioned in Chapter 2, was established in 2000 as a community of people. The company has an online platform where members of Threadless can upload their own t-shirt designs. The members vote, and the most successful t-shirt design is put into production by the company.

Designer story

In fashion design and architecture there is a tradition of designers developing a strong individual signature. If the designer represents a story or a well-known name, this is also used to construct the story about the product and the designer. As we saw in Chapter 2, Georg Jensen and Alessi both use designers' signatures to develop unique products. The story of the designer is embedded in the product and reinforced in the communication of the product.

Personality

The overall look and feel of a product can be recognised as the product's personality traits. Some brands have strongly defined personality traits (Aaker 1997) or brand archetypes (Mark and Pearson 2001). As discussed in Chapter 2, Nike could be understood as an individualistic hero archetype with its slogans 'You never win silver, you lose gold' or 'Just do it'. This personality of the individualistic achiever is communicated through advertisements and is very strong in Nike's products.

It is interesting to see that some brands borrow the look and feel of one category and manage to establish a new identity or strong personality in a new product category. The clothing and footwear brand Cat licenses the brand from Caterpillar Inc., the world's largest producer of heavy construction machinery. By bringing in not only the logo type but the philosophy and the look and feel from heavy equipment, Cat footwear has created a unique identity within its segment.

Ethical considerations

Ethical considerations of how the product is manufactured or how companies behave are also relevant. The shoe brand Toms slogan is 'One for one'. The concept is that if you buy a product from Toms, the company will give a pair of shoes to someone in need. This is not communicated as part of the product except on the product tag. However, Toms shoes have a distinct look and feel that will trigger the brand story if the consumer is acquainted with it.

Origin

The origin of the product may be important. What the country of origin is known for will differ from category to category (Aaker 1996). 'Made in Italy' may be a quality sign in leather and shoes and so-called 'designer furniture'. 'Made in Germany' would have similar connotations in high technology and engineering. With the trend for outsourcing of manufacture, the value of where a product is produced will increase. Whitecollar is a luxurious Chinese fashion brand. The company promises its customers high-end luxury goods. Whitecollar garments are therefore produced in the country with the highest credibility for the particular type of clothing. One such item is a hand-painted silk dress made in Italy – and sold to affluent customers in China.

CULTURALLY DRIVEN STORIES

Brands are important because they engage people and are a significant aspect of culture. Companies are increasingly looking at culturally driven brand strategies to achieve their goals. The role of the products in building the stories of these brands varies. The product may be used as an active communicator of the brand story, but it often serves a symbolic or iconic role.

Relationship

A special status among consumers can provide very strong foundations for how to communicate the product and brand. The English food spread Marmite, made from yeast extract, has a distinctive taste which people tend to either love or hate. In this brand the relationship the consumer has with the product is an important ingredient in the brand story. Marmite knows its customers very well, and uses this love–hate relationship as an important element of its brand communication.

Culturally significant brands

Both fashion brands and brands that appear iconic can have a culturally significant role. How the product is used in these culturally significant brands will vary. In the fashion industry, the products themselves are the core medium for communicating this brand story along with public relations. Advertising is only one aspect of branding in the fashion industry. The theme that is selected for a fashion collection has relevance as an ongoing trend or a driver of change in society, and the products are designed accordingly. Later, this theme will be what dominates the catwalk or other forms of communication.

Brands such as Coca-Cola and Harley-Davidson manage to remain relevant because they employ dynamic communication that is aligned with what consumers care about. The product will play a role in building a culturally significant brand communication strategy. New advertising strategies can help a company to maintain a brand's relevance, although the product itself may remain the same.

Brands with a culturally significant role are also interesting from a product point of view, particularly as the products of these brands often have gained an iconic status. An example of this is the English shoe brand Dr Martens, as mentioned in Chapter 1. In the case of Dr Martens, it was not the company behind the product but people outside the company who constructed the 'myth' around Dr Martens in the first place. In these culturally significant brands the product has changed its role from being a simple product to becoming the very strongest identifier of the brand.

Defining What May Be Coherent

THE IMPORTANCE OF COHERENCE

This chapter has offered a wide range of examples of how products can play an important role in building a brand by performing the brand story. The success of many of the brands discussed derives from the fact that they have found their own version of a story that is distinct for that brand, and that the product is integrated into telling this story. A story like 'ergonomic' can be told in many ways. Presenting an opportunity for a strong product story such as 'ergonomic' to thirty designers will necessarily result in thirty different interpretations.

The company can manage this by carefully considering the story and how it is expressed through the product.

What these brands have managed to build is coherence over time. Coherence is an important pillar in building a brand. It implies that there is some coherence in how the company expresses itself throughout the consumer journey, as described above. This can be achieved through multiple touchpoints, of which the product is just one. How the product represents the brand will need to be consistent – both in the production and presentation of a single product and across different products belonging to the same brand.

One of the important reasons to establish a brand used to be to promise consistent quality. Customers have some expectations about how a product should perform from their previous experiences with the same brand. The perception of quality may also be one of the 'rational' reasons customers cite when explaining their choice of an expensive item from a luxurious fashion brand. If the product then fails to live up to this promise, the perception of what the brand is about may change.

If clothes from a cheap brand wear out quickly, it is still living up to its reputation. If the perceived quality is low, an unexpectedly extended lifetime of a garment will lead to a positive experience for the consumer. On the other hand, if the perceived quality is high and the garment wears out quickly, this can be considered a product failure, which will compromise the brand over time. The experience of purchasing a piece of clothing with that brand diminishes. What is important in a branding context is not always a question of promising a good experience, but delivering to the standard you have promised. Consistent product quality over time is what a brand promises and what consumers expect.

People do not necessarily buy brands for rational reasons, but they often like to use rational arguments when explaining to others why they feel a brand appeals to them. As consumers' communications about the brand are so crucial in building it, their experience of the product is an important aspect of what people talk about when they discuss that brand. This becomes part of the individual's contribution to building the brand. The product thus has the potential to release the expectations customers have about the brand. It also allows people to build the brand based on their own premises, as their interaction with the product cannot be controlled fully by the company.

DEFINING THE PRODUCT DNA

If the company has succeeded in building coherence over time, it may have developed a certain product DNA – a set of characteristics that makes the product unique to a specific brand.

Criteria to Consider When Defining the Product DNA

- The product's role in performing the brand story – level of integration in the brand story, and how to achieve it
- Explicit references – references to the brand, brand signatures
- Implicit references – personality traits, emotional associations
- Unique characteristics and benefits
- What the product should not be – differentiation
- Quality – level of quality, and how to achieve it
- Senses – which are important to think about, and how to stimulate them

---

In defining the product DNA, the company will have to understand how the product is building the brand. The design team will have to go deeper into what this story means in context.

In any product, there are many stories that could be read into it. The product tells a story about the brand, about the production techniques, and about trends that the product is referencing. The product often has multiple layers that will have different connotations for different observers.

In the process of understanding the product DNA, it is important to deconstruct the product and look at what other meanings the same product will communicate beyond what the design team intends. The product can be deconstructed into silhouette, main form, details, finish and materials. The product can also be deconstructed into explicit brand elements that can be called 'brand signatures'. In the packaging industry, product graphics and the text on the packaging will play a major role. In other categories, such as in high technology, the deconstructing process should also include the graphical user interface, as this is linked with how to understand the tangible product.

Both the underlying design principles and explicit and implicit references to the brand are defined in the product (Karjalainen 2004). The explicit references are solid, while the implicit references are part of building the look and feel. It is as important to pay attention to these implicit references as it is to the explicit ones. The yellow stitches on a Dr Martens shoe or the air sole are some of the associations consumers have with the brand Dr Martens today. By being consistent in always including these in new shoes that are developed, these have over time become an explicit reference to the brand Dr Martens.

Implicit references are by definition not clearly stated, and are therefore difficult to define. It is possible to gain a sense of what the implicit references are by using personality traits or emotional characteristics. Dr Martens shoes can, for example, be described as 'rustic' or 'rough'. In defining the product DNA, it can be helpful to look at similar products and try to explain why these are different. Identifying what *is not* part of the product DNA is as important in product design as knowing what *is* part of it.

By defining the product DNA and using this as a tool for making decisions, the design team can take one step further towards developing strong coherence in the brand. This is important in planning a coherent product portfolio and developing future products.

DECONSTRUCTING THE PRODUCT

The product as a communicator of meaning has been an important part of the design agenda since the 1980s. Understanding, constructing and deconstructing meaning relies on semiotics, the science of signs.

There are two directions in semiotics. The Swiss linguist Ferdinand de Saussure's models identify two components that make up a sign: the concept of the word (the signifier) and the sound that signals this word (the signified). The American semiotic philosopher Charles Peirce's model was a three-component system which also included the interpretation of the word and the sound (Chandler 2007). Using his theories may reveal the relationship between the design team's intention, the product, and how the user understands the product (Vihma 1995).

This thinking can be translated into the interaction between the message (the idea behind the product, often rational and emotional benefits), the physical product (representation of the message) and the interpretation (the consumer's understanding of the product) (see Figure 3.8).

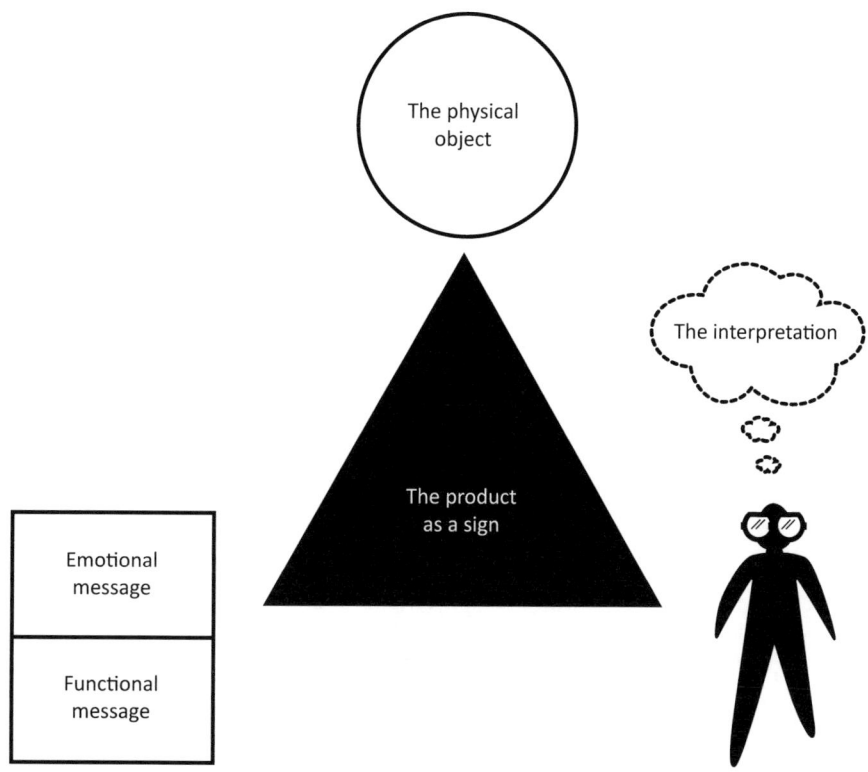

Figure 3.8   The message, the physical object and the interpretation

The design team's message may be both rational and emotional. For example, the design team may want to design a radio that is easy to use (rational message) and one that also evokes nostalgia (emotional message). This might be translated into the form of buttons that reference buttons on an old radio. The buttons might be concave to signal 'push', the references to the radio giving the form an old-fashioned look and feel.

What the product communicates depends on the person interpreting the message, as well as the context in which the product is presented. For this reason, in order to understand the meaning of the product, the context should also be included in the analysis (Vihma 1995). Various factors influence how a user interprets the product. Each user will have a background that influences how they understand the product. The culture the user is from, the context in which the product is presented, the user's educational background, the user's family background – all of these will affect how the user understands the meaning. The design team may have a message it would like to communicate through the product or it would like the product to represent. However, it is only when someone – in this scenario, the consumer – interprets the product that its meaning will be communicated. In brand-building, it is important to be aware of the distinction between how the design team defines the object and what it means, and how the consumer understands it. This means that the product or elements of the product do not become meaningful brand representations unless the consumer understands them in the way the design team intended. According to the design theorist John Heskett, there is a need to understand the 'interplay between designers' intentions and users' needs and perceptions. It is in the interface of the two that meaning and significance in design are created' (Heskett 2002: 54).

There will inevitably be distortion in the process of defining the meaning and translating it into a designed object, as well as in how the consumer understands the designed object (Karjalainen 2004) (see Figure 3.9). The design team may work on the basis of an idea about the brand story or a new opportunity it has identified. The idea the product represents is usually first described in words and a few pictures. These will be translated by the design team into a tangible object – a product. How the design team understands the briefing will direct the translation. This is the first distortion in the communication process. Later, the user's interpretation of the product will depend on various factors. The distortion created will result from the consumer's socio-cultural background, previous experience, knowledge about the product, education, as well as the mood of the consumer at the time of use. Distortion is also created through the way the product is presented. The above-mentioned bottle from J.P. Chenet may be interpreted as referencing old production techniques if the consumer has previous experience with glass. Other suggested interpretations may be that the asymmetrical shape references a drunken person or that it references use.

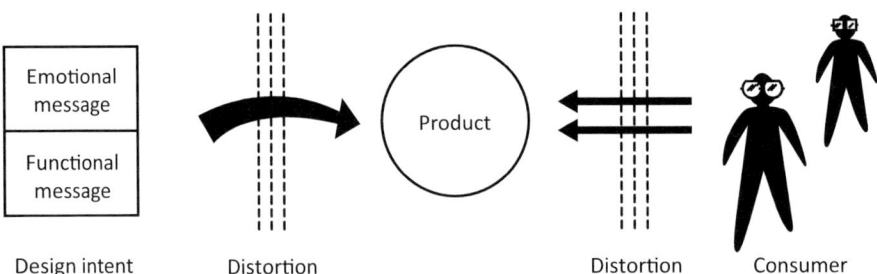

Figure 3.9   Distortion in semantic transformation
Based on Toni-Matti Karjalainen's model of semantic transformation.

Brands with Substance

The aim of this chapter was to investigate how a product can play its part in building a brand. How a product represents a brand will depend on the chosen narrative and the product's role in communicating this narrative. In strong brands the product plays an important role in performing the story of the brand. Both what the product looks like and how the product behaves in use are important aspects of building the brand and releasing the brand promise. The product is what makes the substance in the brand story. How the consumer perceives the brand is based on the overall experience, extending from initial campaigns, to use, and further to disposal or reuse. Over time the company will have developed certain characteristics that are associated with its products – the product DNA. Understanding the context the product is part of is of key importance in defining the product DNA.

Context: A Dynamic Learning Process

The product development in a company is a result of a co-operation where the designer is only one of many parties involved. When I design a product I want it to hit the shelves the way it was intended. To achieve this, I have to get the buy-in of the board of directors, the product development manager, the engineers and the marketing department. To make this happen, everyone needs to have the same basic understanding of what the product is, what the brand is, and what the product does for the brand. If you isolate yourself you have a problem in getting your ideas through.

Cathrine Movold, Designer and Strategy Adviser

This chapter will place the product design process within the larger context of building the brand. Product design is only one of many activities a company undertakes that will have an impact on how a product is perceived. Other activities, such as promoting the product or appointing a new leader of the organisation, are also important in branding. These other activities outside the product design department will also affect the product design process.

Design and Branding: An Integrated Process

The Norwegian oral care company Jordan (branded as Wisdom in the UK market) has placed a particular focus on design (Hestad 2008). The director of its Nordic home market also highlighted design as part of its strategic competitive advantage in the oral hygiene segment. His definition of 'design' was vague, but it was clear that it was an integrated part of the company's branding. The Jordan oral hygiene brand DNA is loosely defined, but the product is a key touch-point with the consumer, and is seen as 'the person' representing the brand

(Wentworth 2007). This touch-point is responsible for delivering the brand promise. The toothbrushes should all be 'relevant', simple and authentic. Another key concern is that the products should be unfussy, with clean surfaces and few gimmicks (Hellerud 2007; Wentworth 2007). The product also becomes the advertisement, which is important in the market in which Jordan operates as the products communicate a story that is easy to understand and attracts consumers. When consumers enter a store, what they see are the products. The rise of powerful retailers and competition in buying shelf positioning makes it even more important for the product to communicate a story consumers find appealing. In this section we will explore the design process behind the Jordan Individual toothbrush, which was released on the market in 2007 (see Figure 4.1).

THE PRODUCT AS AN IMPORTANT TOUCH-POINT

In 2005 Geir Hellerud, Jordan's Product Development Manager, was reading an article about the Swedish design company Ergonomidesign in a Scandinavian Airlines (SAS) flight magazine (Hellerud 2007). Ergonomidesign was founded in 1969, and has built its reputation on ergonomics in the design process. Hellerud saw this as a good opportunity to develop a Jordan toothbrush that was optimal in an ergonomic sense. This was the starting point of a new design process – a potentially strong product story that it was possible to sell. In Jordan's case, the story is usually designed by either the Category Development Manager (from Marketing) or the Product Development Manager (from Research and Development). Where the idea comes from varies, but they often seize on a type of customer behaviour they have identified as a potential opportunity for a toothbrush. This opportunity could be a new behaviour that has been observed, a trend or a story that has been communicated in another product category. In this case it was the opportunity to make toothbrushes with a story about being ergonomic.

Jordan first invited Ergonomidesign to conduct an ergonomic study on toothbrushes. The study identified two groups of brushers: 'scrubbers', who have a solid grip when brushing, and 'cleaners', who have a delicate grip. Based on this project, they decided to go forward to design a product based on the opportunity to launch an ergonomic toothbrush.

As soon as the opportunity had been defined and explored more deeply, the design team wrote a design brief that was given to two design agencies. What is presented in the design brief will change according to the project.

Figure 4.1   The Jordan Individual toothbrush, designed by Geir Øxseth
for Jordan (2006)

Photo: Stine Heilman, Copyright Jordan AS.

The professional background of the authors of the brief, who the company would like to work with, and which opportunities the company is seizing will influence how the brief is written. The design brief specifications for Jordan included: 'Scandinavian look and feel', 'ergonomic toothbrushes', 'relevant to the market' and 'design'.

An important factor in how Jordan develops its projects has been to invite two different agencies to compete for the brief. Jordan then selects the designer based on whether it has already worked with the designer, or on which it thinks will be more suitable for a particular brief. For this brief Jordan decided to invite Ergonomidesign as it had been involved in the first phase of the project. Jordan also invited Geir Øxseth, who had been working with Jordan for years and was familiar with its processes. He had also designed the Jordan Trend dishwashing brush, which had received much praise when it was launched on the Norwegian and British markets in 2000 (in the latter market by the company Addis).

Figure 4.2   A range of Jordan Individual toothbrushes, designed by Geir Øxseth for Jordan (2006)

Photo: Stine Heilman, Copyright Jordan AS.

The team at Jordan followed the design process until it had developed prototypes for two brushes. The team was as interested in what did *not* correspond with the brand as what did. The toothbrush concept Ergonomidesign had developed to communicate 'ergonomic' appeared too aggressive for the Jordan brand. Øxseth's toothbrush was in line with what Jordan felt was its look and feel. One of the key arguments for selecting the Øxseth design was simply that it was a beautiful toothbrush. The brush also allowed for a wider range of graphics to be integrated into it. Opening up this possibility changed the story to be communicated. The focus shifted from telling a story about 'ergonomics' to telling a story about 'individuality'.

The Øxseth design also included an element that allowed for a greater range of individuality (see Figure 4.2). Individuality was not necessarily a question of having the right handle or anything about the brushes themselves, but about the possibility of including different graphics on the product according to one's own taste. Outsourcing production to Malaysia allowed greater flexibility in the design process. A new moulding technique enabled a multi-layered process that made it affordable to incorporate a range of graphics on the same basic brush handle design. This turned out to be a far more complicated process than first imagined, but after several trials and errors the new design met all the requirements.

In the first release of the product, Jordan introduced two different handle sizes based on the identified groups of 'cleaners' and 'scrubbers', along with two different softness levels of brush. They also introduced 16 different graphics incorporated into the toothbrush handles. At first the team at Jordan was not sure how the complexity of the message would be conveyed by introducing such variety, but the toothbrushes quickly became popular in the market (Wentworth 2007). The option of changing the graphics made it possible to redesign the toothbrush without having to redesign the handle, thus extending the product's longevity in the market. The second launch of the toothbrush introduced a wider range of graphics. In the third launch Jordan invited consumers to compete to have their own graphics printed on the handle. The fourth launch invited well-known designers to design the graphic, making the brush a 'designer toothbrush'. In a recent re-launch, the campaign was driven by a marketing company which invited consumers to send in their own styles. The meant that the product could be re-launched with a new story every year without having to start a new design process.

The Jordan Individual case shows how the design process may begin with identifying an opportunity in the market, but may change during the design process. The conceptual story that it may seem possible to communicate in marketing material may change in character when translated from words to a physical form. How this idea will apply to the brand and be part of building the desired look and feel depends on who the designer is and how this idea is interpreted. Although Jordan's toothbrushes focus on being ergonomic, the style Ergonomidesign initially presented to communicate 'ergonomic' did not fit with the look and feel of the Jordan brand. The design process therefore involved the company learning about its brand while designing the product. By designing products with a strong story that it was possible to follow up in the product's promotion, the company managed to build a strong brand that is relevant for customers.

DESIGN AS AN INTEGRATED EFFORT

The Jordan Individual toothbrush is a good example of how seeing product design as an important aspect of branding can create a unique and differentiated product in the market. 'Design' was defined as an important strategic advantage, and was followed up by key people in the process. In this case, stakeholders both inside and outside the organisation had roles in defining the product. The process changed depending on how the designers interpreted the brief and what they identified as a potential market opportunity. The idea of the product, its newness, and how it relates to current trends and competitors' products all need to be taken into account in deciding how the product can build the brand. In this way, the aim of the product design process was not only to provide a product to meet a specific need, but also to be a promotion exercise. The key message in the promotion of the product was also the opportunity that the design process started with.

To maximise the effect of design and branding it is important to see these as integrated activities in the overall company strategy. This will also include bringing in design early in the product development process. If the design team is brought in at the very end of the process, there will most likely be strong limitations on what it is possible to do. The result may be a 'quick fix' to 'add some differentiation elements'. This may be a cheap solution for the company in the short term, but it may also lead to a cheap appearance, and not necessarily a product that is representative of the brand. Consideration of how the product can build the brand is not an activity that can be brought in at the end of the product design process, but must drive the whole process.

AN INTEGRATED DESIGN AND BRANDING PROCESS

In brand-building it is important to have both an 'inside perspective' and an 'outside perspective' (Ind and Bjerke 2006). An 'inside perspective' in this case refers to developing the brand value proposition from the ideas or values of members of the organisation. A brand value proposition is what the brand proposes to the consumer as being valuable. An 'outside perspective' means developing what to propose to the consumer by positioning the brand according to other actors in the market. This could also include bringing in insights from trends and the culture the consumers are part of. Brands that manage to offer value to the consumer through something that stems from the company's own efforts while at the same time establishing a position that is unique emphasise their distinctness. In the Jordan case presented above, both the inside and the outside perspectives were important. The company was defining the opportunity and had its 'guardians' in place to guide the design process. The designers it brought in were granted freedom in how they designed the product. Therefore, the Jordan case represents a dynamic interplay between various stakeholders from inside and outside the organisation.

How the design process integrates with building the brand can be summarised in a Brand Activity Framework, as shown in Figure 4.3. This framework consists of several stages, the design process being one of them. The other stages are: brand DNA, translation, aligning touch-points, making the product available to consumers and monitoring use. The Brand Activity Framework consists of three layers: the company, the market and society. 'The market' refers to what is traditionally seen as the market space and actors such as retailers, competitors and interaction with customers. This is where the product is made available and sold to the consumers. The market is the transition to society. 'Society' consists of customers' daily use and their interaction with the product as well as with each other.

The Brand Activity Framework based on the Jordan case represents one way of seeing the design process as part of the branding process. Other business models may change the dynamic. If the company relies on external consultancies in the process of designing and promoting its products, the possibilities and limitations will differ from those for a company that has a large in-house design team. Facebook was mentioned in Chapter 1. For such a business, the model will look different, as it is not evident what the 'market' and 'society' are for Facebook. Organisations that have managed to build a brand have done so because of their ability to define a successful business model that is unique. The Brand Activity Framework will therefore have to be adapted to fit the current business.

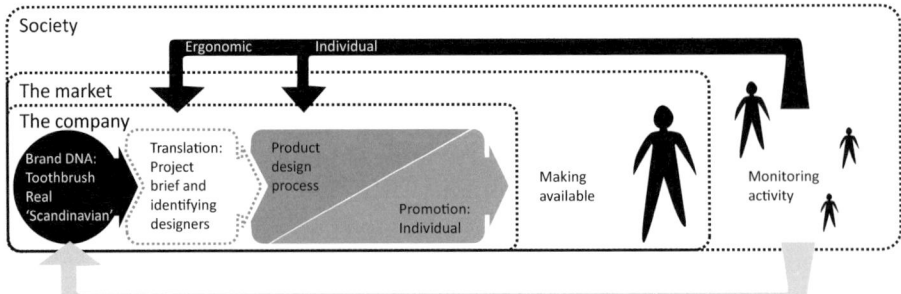

Figure 4.3   The Jordan Individual toothbrush design process – Brand
Activity Framework (author's reconstruction of Jordan process)

Principles in a Design-driven Branding Process

BRAND DNA: VALUES AND HERITAGE

In the framework presented in the Jordan case (see Figure 4.3) the brand-
building process starts with what is called the brand DNA. Ideally, the brand
DNA will underlie every activity in the company, and in this model it is defined
by two components: core values and heritage.

The core brand values have been described as the heart of the organisation
(Aaker 1996). In most companies the brand values will be created by the
culture the company has developed. It will often have been the founder of the
company who first defined the organisation's values, but these values need to
be maintained and communicated to the organisation in order to have an effect.

The global consulting engineering company Arup is an example of
successful communication of what it defines as the company's values. The
founder of the company, Sir Ove Nyquist Arup, set out these core values in his
'Key Speech' (Arup 1970):

- We will ensure that the Arup name is always associated with
quality.

- We will act honestly and fairly in dealings with our staff and others.

- We will enhance prosperity for all Arup staff.

The key speech was introduced 24 years after the company was founded, but at a time when it had grown to a size where definition of its values was necessary. In 1970 the company was in the middle of its first big international project, the structural design of the Sydney Opera House, and would see rapid growth to its present 10,000 staff globally. As a smaller company, everyone would have known the founder personally and a definition of the values might not have been necessary, but in 1970 Arup had seen the start of its growth and the founder had turned 75 years old, thus creating a need to write down the company ethos.

Another element that is important in understanding brand DNA is heritage. Established brands will have a heritage, and all activities concerning the brand add to this. Brand heritage in relation to brand DNA is therefore understood as past activity. Brands such as the English clothing label Barbour are labelled 'heritage brands'. Their heritage is important in telling the brand story. However, all brands have a heritage. Activities a company is involved in become part of its heritage. Arup saw a project like the Sydney Opera House as forming a major part of its heritage (Jones 2006).

Brand DNA will be shaped dynamically by what is happening, both in the market and in society. If the market has responded negatively to former products, the company will most likely have to change its brand DNA. This will be explored further in Chapter 6, which presents a case where changes in the market situation led to changes in the core values of a brand.

CHANGES IN THE ORGANISATION

All activities in the organisation could affect the brand DNA, but change of leadership is recognised as a force that can change the whole concept of a brand (Aaker and Joachimsthaler 2000). Fresh leadership can also alter the design of the product. When Nokia brought in a new leader, Stephen Elop, from Microsoft in 2010, the design of its phones changed. The aim of this was to bring back Nokia's past glory. Damian Mycroft, who was a Senior Design Manager in Nokia at the time, explains:

> there was this massive change – having the design language mean something about Nokia. The design language was more progressive, it was more human. So when they made the announcement about Nokia and Microsoft, there were renderings created in the design team. The question asked was what the brand Nokia looks like as a product.
>
> (Mycroft 2011)

When an organisation changes its vision, as in the example of Nokia, this will have a major impact on the design process's role in building the brand. A drastic change will open up new approaches and innovations in the product design department. The values and the vision should be a source of inspiration for the designers, while at the same time building the brand in a coherent fashion. Damian Mycroft elaborates: 'If you manage to communicate coherently, but have a set of values that are broad and high enough – that is a good combination to be able to do new things and bring people along with you' (Mycroft 2011).

Large organisations with many stakeholders can be particularly challenging when seeking to form these visions. It is necessary to manage stakeholders and ensure that the appropriate input is solicited for key parts of the design process. If not, there is a danger that the design will not reflect the collective vision of the company. According to Paul Marchant, Head of Product Design at Transport for London, some of the most enduring designs on London Underground came about in the 1930s, when one person could have a vision that a whole company could buy into:

> Some of the most fundamental change, and as a result enduring design, was led by Lord Ashfield.[1] He created a clear vision and strategy that the people bought into. What you have now is a kind of democracy of opinion which is subjective rather than objective.
>
> (Marchant 2011)

The larger the organisation, the more challenging it seems to be to bring everyone on board and have them take ownership of the vision. Design that is led by committee without a clear vision tends to result in compromises and the lack of a coherent product with a strong story to tell in the market. It is important to involve a large proportion of the organisation's staff in the design process, to create transparency and promote sharing of information. There is a need for a strong product story, a clear direction and a vision that all members can buy into. It is equally important to define who is in charge of decision-making.

A common feature of companies that are successful in using the product design process as part of building their brands is that their leaders have an appreciation for design and an understanding of how design is a strategic

---

1   Albert Stanley (Lord Ashfield from 1920 onwards) was Chairman of the Underground Electric Railways Company of London (UERL) from 1910 to 1933 and Chairman of the London Passenger Transport Board (LPTB) from 1933 to 1947.

resource for their organisations. To develop a culture that fosters design, the understanding of the product's contribution to building the brand needs to be nurtured over years. Such companies have people in prominent positions, either in middle management or top management, who act as ambassadors for design. These ambassadors have the power to encourage other members of staff. If leaders do not have an appreciation of design or an interest in it, the design department will struggle to be allowed to participate early enough in the process. One of the respondents in the research for this book said that he had to force the organisation to let the design department take on a role in the development process. Pushing strongly to include the design department will undoubtedly create tensions, but it is essential in developing an integrated development process.

In another company the design department started with its own project to inspire the organisation to use design as a key competitive advantage. Initially, the project manager felt that he had to keep the development process a secret from senior management as they had little interest in product innovation. However, when the managers saw the prototype of the new design, they could grasp its importance, and the project led to a fundamental change in the company. Making a prototype meant that all members of the organisation could understand and share the same vision. Prototyping is an important and powerful way to make a vision real and tangible, and in this case the prototype was what managers needed to see in order to believe in a change of the company's portfolio. This later changed the brand DNA of the company.

MAKING BRAND DNA AND STRATEGIC DIRECTIONS DESIGNABLE

How the brand DNA informs the design process will differ depending on the structure of the company as well as how the brand DNA is defined and what it represents for the company. Is it the company name that has become the brand, or is the company the 'mother brand' behind several sub-brands? Is the company recognised by a particular concept or approach in what it delivers? Do former products create a strong heritage that needs to be maintained? The relationship between the brand DNA and strategic directions that are designable will need to be clarified.

As Sir Ove Arup mentioned in his 'Key Speech' (1970), the core values of Arup are commitment to quality, an honest attitude towards clients and staff, and enhancing employees' wellbeing. These are important in building the Arup brand, and are apparent at the strategic level of the individual projects the company undertakes. However, they will be too abstract to be useful on a day-to-day basis to direct projects. The brand DNA will need to be translated into design principles or guidelines that can inform the work undertaken by the company.

In the Jordan case, dental care was one of three departments in the company (the others being cleaning and painting tools). The core values that Jordan claims to represent in their brand DNA handbook were too abstract to be used in the development of a brand in the oral hygiene category. The dental care department therefore defined a set of criteria or principles internally that could be understood as Jordan's oral care brand DNA. These were guiding each of the processes. The design process in the Jordan case started with identifying an opportunity that could become a strong product story. In the brief, and by guiding the designers, the team took an active part in translating and revitalising the brand DNA. Jordan's heritage defined how business had been done earlier, and gave clues to how customers might experience new products. However, the brand DNA seldom informs the design process directly. In the Jordan case, the core values were seen as guiding the company's activities, thereby indirectly shaping the product.

In the translation process the design team will have to be open to changes internally as well as externally. The example of Nokia illustrate that a major change a company undertakes may mean that the brand DNA will have to be reformulated. On the other hand, there will be constant changes in society which will have an effect on how the brand is perceived. The brand DNA will have to be translated according to contemporary trends and according to values or directions that are possible to articulate in the design process.

THE IMPORTANCE OF ALIGNING TOUCH-POINTS

The Brand Activity Framework comprises all the activities of a number of stakeholders. All of these stakeholders' activities affect the development of the brand and how it is perceived. The complexity of this framework makes it crucial to develop a strong culture in the organisation. This is also why it is important for the organisation to narrow the scope of what it seeks to build as the brand. Clarity about the brand internally will help to engage people in developing the brand, and can be important in creating coherence across all the company's activities. The Brand Activity Framework outlined in Figure 4.3 depends on aligning the activities within the company. The organisation will have to establish coherence in what it wants the brand to be about. It will also have to know what the product is about and how it should build the brand. The product will have to be aligned with the intended brand story. The story that is driving the design process should also be reinforced in related advertising. Promotion of the company and the product are well-recognised tools to develop and strengthen brand identity, and are widely explored in the literature. By aligning the brand, the product and advertising, the brand will be recognised as having a strong identity. The company's communication strategy should be aligned with how the company would like the product and the brand to be perceived. In Figure 4.3 the promotion of Jordan Individual begins when the product development starts.

THE START OF THE NEW PROCESS

The process of building a strong brand is not linear. It is a dynamic process. All the activities in the company, the market and society are important factors in building the brand. Consumers' experiences with the product will affect how they perceive the brand, which again refers back to the brand heritage of the company. Therefore, making the product available to the consumer is not the end of the process – monitoring the use of the product as well as changes in society will become the beginning of the next design process.

The real learning of how the product could build the brand will come when the product is launched on the market and people start to interact with it. Information of how people respond and interact with the product will need be fed back to the brand DNA, and will be part of shaping the future development process. It is not only the product development process that will need to be informed, it is all the stages – defining the brand DNA, translating it, the design process and aligning touch-points. These stages will draw on learning from past activities, in the company, in the market and in society.

Three Layers of Uncertainty

The three aspects – the company, the market and society – represent three layers of uncertainty in the brand-building and design process, as only part of the process can be controlled. Depending on what kind of company it is and which category the company operates in, there will be limitations and opportunities connected to the layers when it is possible to integrate the product design process as part of building the brand. Company heritage, the structure of the organisation and the current context are just a few of the pillars that will define the context.

SILOS IN THE COMPANY AS BARRIERS TO COHERENCE

The company context is where the highest degree of control and management is possible. Within the company, everything from the budget to the people in management positions to the designers will play its part in defining how the product design process serves to build the brand. Outsourcing the design work or manufacturing, as many companies are doing, will add another level of uncertainty. Even though there are massive uncertainties in the market and society, some of the most difficult challenges will arise within the company itself. The larger the scale of the organisation and the more complicated the product portfolio, the more challenging it seems to be to use the product design process to help to build the brand. There are many internal issues that will need to be resolved. One of the key challenges managers face is that of so-called silos within the company – where each department has its own agenda, and these sometimes conflict with each other. This lack of communication between departments was recognised by many of the professionals interviewed in the research for this book.

If the product development department, the marketing department and the branding department in a single company do not communicate with each other, the collective contribution of the company will be a fragmented brand. Without strong integration between these departments, the design process will most likely fail to build a coherent brand.

One such example was experienced by designers working in a large consumer electronics company. The consumer-facing units saw that consumers were beginning to be more interested in a basic lifestyle, and defined this as one of the key areas they wanted the brand to signify. In consumer electronics the product plays an important role in how the brand is experienced. However, the design department experienced being left out at the start of the process of changing the brand DNA, and the message of creating basic and easy-to-use products was only communicated in the marketing material. As one of the designers framed it: 'The key challenge in the company was to get the marketers to understand that, when building a brand about a basic lifestyle, the products need to deliver on this promise.' As a response to this the design unit started its own project and invited marketing to take part. However, the business units in charge of manufacturing and bringing the products to market were left out of the equation. The rationale behind this was to ensure free thinking and innovation, without the constraints of the strategies the manufacturing and distribution units had already drafted. At the end of the project the conceptual products were introduced to the manufacturing and distribution units, but due to the lack of transparency there was little buy-in. Only very few products ended up in the pipeline and were produced. Failing to establish a design process that included manufacturing and distribution meant that the company did not succeed as it could have if these shortcomings had been addressed.

Many subtle decisions that will have an effect on how the product is perceived are made during the design process. The existence of silos was also mentioned as one of the key challenges in maintaining a coherent look and feel. If decisions on CMF – colour, materials and finish – are made in one department while the form of the product is decided in another, it will be necessary to have someone in charge of the overall look and feel. The Jordan case exemplified how important it is to have a small team, or a single manager, in charge of the design internally. This team will have to be able to work across departments. It will need to have a strong vision of what the product could be, and how the product will represent the brand. Only by following up all the involved business units in the company can the product become a strong ambassador for the brand. We will return to this issue in Chapter 5.

REAL LEARNING COMES WHEN THE PRODUCT IS LAUNCHED

How the product contributes to brand-building after it has been launched will largely be beyond the company's control. In the market context the company has some control, but not to the same degree as in the company context. In this space there are various stakeholders who will have an impact on how the product is perceived. Nevertheless, it is partly within the control of the company as it can influence a number of aspects of the market that will affect the brand. The company can to a certain degree choose which store, what kind of distribution and which marketing channels to use, and in some cases how to present the product within these channels. The level of freedom depends on the status of the product and the company's budget. A market leader with a larger budget will experience more freedom than an underdog trying to break into a market. The retail space is often controlled by the organisations behind the retail chains, and another level of uncertainty is competitors. In a retail space the products are displayed in the categories they belong to, which means that products are lined up next to their competitors. The products are understood as belonging to the specific category, perhaps through being sold in a particular store or being distributed in a particular way.

This was one of the lessons Jordan learned when launching another toothbrush, the Jordan Go! It seized what it saw as an opportunity involving the 'third toothbrushing'. Toothbrush users usually brushed their teeth in the mornings and evenings. 'On-the-go' people used chewing gum or mints to freshen their breath. The idea was that Jordan had an opportunity to introduce a new product which would not impinge on other toothbrushes in the market if it could develop a toothbrush that was perfect for use 'on the go'. This led to the Jordan Go! design project, for which the firm Formel Industridesign won the brief. The toothbrush was designed in a pocket format. The toothpaste was formulated so that it could be swallowed and was integrated into the unit, and the brush was softer to prevent damage to the teeth that can sometimes occur through over-brushing. However, when launched on the market the product communications focused on travelling and everyday situations when it is handy to have access to a toothbrush. The product soon became popular as a travelling toothbrush, but the idea of a third toothbrushing session was not well received.

There may be many reasons why this product did not succeed as planned. The main one is probably that it faced the challenge of changing people's behaviour. Achieving this would have required a more focused effort. The communication also appeared confusing, as Jordan was promoting multiple scenarios for use, and users did not see the toothbrush as a replacement for chewing gum.

If a company introduces a new product to the market, with a new user scenario, this will need to be communicated explicitly. In the communication for Jordan Go! the messages presented both a travelling toothbrush and a toothbrush for convenient use 'on the go'. Changing user behaviour can be time-consuming, and the company will have to be patient. For Jordan, the toothbrush met with such success as a travelling toothbrush that it changed the product and the communication to make for an even better fit with this scenario.

The core activity of the market is to make a product available to potential buyers. The process of making a product available to customers is an important consideration in building the brand. New meanings will be added to the product depending on the choice of distribution, market channels, environment and the price of the product. All of these are key elements in the planning the company undertakes. However, these factors are also greatly influenced by stakeholders outside the company. Chapter 7 will present a case study of the

Norwegian company Stokke, a brand in the children's product segment. Prior to adapting the entire business, the company was facing a major change in the retail situation. In the children's product segment there had been an increase in strong retailers. Stokke originally had only one product in the segment, while retailers preferred to carry several products from the same company. There is a similar situation in the fast-moving consumer goods categories. Companies have to pitch their products to these retailers – if the story is not strong enough, the product will most likely not be accepted.

NEW LAYERS OF MEANING

When the product is bought it becomes part of the wider context of consumer culture, and many companies will indeed be surprised by the use of their products. It is important for a company to understand how its products are perceived and become part of this context, as this is where the brand promise is realised.

In using the product, consumers are adding new layers of meaning to it, and certain rituals have been identified that users perform when using products. The product can also change meaning through interplay with the user.

The American anthropologist Grant McCracken (1990) explored how users had several rituals whereby they added meaning to a product. Gifting was one of these rituals. When a person receives a gift from someone else there are often some thoughts behind the selection of a particular product. Also, the process of gift wrapping the product adds new layers of meaning to it. Another ritual is grooming. When dressing for different occasions, accessories and hairstyle add new meaning to clothes. By using the product in relation to other products, the product changes meaning. A third ritual is when a customer buys a second-hand product. The product may be cleaned or some accessories carefully selected to go with clothing. This is part of transforming that product from being someone else's to becoming the belonging of the new owner. The fourth ritual McCracken explored was the process of buying new clothes, and making those garments part of other products that belonged to the owner. All of these are rituals that people perform, and that contribute to adding value to the product.

If a product is picked up by celebrities, it can provide additional value for the company behind the product, due to the celebrity flair associated with it. This is called 'celebrity endorsement', and it has been used actively in building certain brands (McCracken 2005). There have also been cases where the opposite situation has applied. The luxury champagne brand Cristal from the champagne house Louis Roederer became a favourite among rap artists. An article in *The Economist*'s summer special in 2006, *Intelligent Life*, was titled 'Bubbles and bling', and dealt with such unwanted attention. In it Roederer's Managing Director, Frédéric Rouzaud, was asked whether Cristal's association with the bling lifestyle could hurt the brand. He replied: 'That's a good question, but what can we do? We can't forbid people from buying it. I'm sure Dom Pérignon or Krug would be delighted to have their business' (*The Economist* 2006). This was not taken lightly by the rapper Jay-Z, who was mentioned in the article. Jay-Z subsequently sent out a press release which stated that the brand was 'racist' and that he would withdraw Cristal from his chain of bars (Arnold 2006). The Cristal case illustrates the impossibility of managing the cultural sphere.

Society cannot be managed, but activities and behaviours can be monitored carefully, and companies can adjust their activities accordingly. In this context new layers of meaning and stories will be added to the product. By systematically monitoring how the brand and the product is perceived can a company decide how to respond to these changes. People's activities on the Internet are an increasingly important source of information for companies (Li and Bernoff 2008). How the product is used, in which settings, and how users perceive it will be important considerations.

DESIGN AS PART OF THE BRAND: A LEARNING PROCESS

In *The Reflective Practitioner* (first published in 1983) the MIT scholar Donald Schön described how professionals think in action (Schön 1983 [2011]). Professionals, he says, reflect in action. He describes how designers think by making many sketches of their ideas. The designer has an opportunity to create a virtual world by using models and sketches that allows for testing of ideas before developing products. This allows designers to make many trials and errors before defining which of the solutions is viable. This is what happens in most design departments: the design team tests many different solutions, and decisions are made consciously and unconsciously while designing. This is a key characteristic of design, and gives a good description of how decisions are taken that eventually lead to the final product.

The process presented in the framework this chapter builds on has some similarities to the individual design processes, even though brand-building is part of the bigger picture. The real learning about how a product efficiently builds a brand will not be available before the product has been launched and people start to use it. In some cases, especially in the case of radically different products, the real learning can only be observed years later. The stakes are much higher when launching a product in the market than when a product is still on the design department's sketch board. Therefore, learning from the lessons when the product is launched is crucial to future product development processes.

It is important to build an organisation that provides room for learning. Many organisations appear to be hostile environments with little room for learning, both from their own experiences and through pushing boundaries. If someone has failed, organisations often seek to allocate blame rather than learning from the experience. Reflecting on both successes and failures is a crucial part of building a sustainable brand.

How the design process will build the brand will vary depending on the context the product is part of. In the design process all of the stages in the product's life cycle can be mapped and planned. This is particularly, and increasingly, important for the stages of disposal or reuse of the product. The better the company plans for these stages, the more control the company can gain over the process.

Learning from the Context

This chapter aimed to describe the product design process as part of a larger context. It introduced a Brand Activity Framework setting out three layers of activity: the company, the market, society. It is important to consider all of these layers to establish how the product design process serves as part of building the brand. Brand-building and the design process have many similarities. They are value-driven, iterative, and needs to be informed by consumers. Every product that is designed and launched to the market gives rise to new lessons for the company and will change what the brand is about.

Although it is obviously possible to manage a company, there are still internal challenges and barriers that may hinder the use of design as a strategic resource. Therefore, seeing the design process as an important aspect of building a brand is a question of management. Only when key decision-makers in the company understand the role of design and build a culture around it can it be harnessed as a strategic resource in building a brand.

Planning: Preparing for the Use of Design Knowledge

All tube trains should be recognisably London: the use of colours, the use of lighting, the train interiors, the moquette to the armrest. The customer information is always set, the line diagram, the tube map. The overall feeling should be calm, and it should be iconic. When you get on a London underground train – you should just know without thinking that it is a London Underground train.
Paul Marchant, Head of Product Design, Transport for London

This chapter will focus on planning and preparation to enable the organisation to make better use of its brand and design knowledge. It will look at how to capture knowledge, define the key team, and how knowledge of the brand DNA can be translated into something on which designs can be based.

Maintaining Heritage as Assets

MAINTAINING THE BRAND

Building a brand is not an activity that can be accomplished overnight. Most organisations spend years building the identity of the company, and only after a while does it become recognised as a proper brand. In this, past projects and what is already recognised as the brand serve as the company's assets. Maintaining what has already been established then becomes an important part of continuing to build the brand. When products are considered an aspect of the branding, past products become part of the brand heritage. The thinking behind these products, how they were developed, and later, how they are perceived in the market form some of the knowledge of how the product builds the brand. Every design project eventually becomes part of the heritage of the company. This heritage will also influence what consumers expect the company to come up with next. The design process behind a product will be an important component in what the company needs to document. This includes the ideas that were developed, but equally importantly, the market situation and the trends it reflects.

Many managers face high turnover of staff, and instilling new staff with an appreciation of the brand heritage can be a difficult and lengthy process. This was highlighted as one of the core challenges by the marketing director in a major fast-moving consumer goods organisation who was interviewed for this book. Turnover is a challenge because new marketing managers and designers are often eager to put their own fingerprints on the products instead of continuing with a design language that builds a brand over time. In this process, it is not helpful that branding and design are both fuzzy, undefined processes that are often maintained by designers' and marketers' intuitive approaches to what is right for the development of the brand or the product. If the knowledge one key person holds is not captured in the organisation, and the rationale behind decisions is not transparent, changes of key people will most likely result in the disappearance of knowledge of what the brand is about.

Failure to maintain the brand through the product was also recognised by one designer who was interviewed, who was surprised by the lack of planning and interest he experienced in maintaining the brand:

> I am surprised to see that the company has managed to build a strong brand with a strong heritage – and that they let so many coincidences affect how the company maintains the brand. This is something I have seen with several brands. You may think that it is a strategy and a plan behind, and smart working over a longer period, while the reality is that randomness plays a major role.

While part of the randomness in this case could be attributed to market changes, the frustration the designer expressed was due to lack of appreciation and understanding of the need for a consistent approach within the company.

In this case, in the end it was an external agency that maintained the brand's heritage. This is not a unique situation. When working internally, it is easy to lose sight of the importance of maintaining what has already been built. Particularly in the case of iconic products or products in FMCG markets, the product itself will seldom go through radical changes. As a designer or a marketer, it is important to recognise that consumers experience the product as just one among many other products. For them it is not a question of changes, but of recognition, trust or past experiences. Established brands therefore need to consider what the company has managed to build successfully, and whether consumers will appreciate change.

On the other hand, change can also be an important factor in building a brand. Organisations frequently fail to maintain their position because they have failed to recognise and adapt to important changes. This opens up positions for competitors, as consumers may have already changed their attitudes because the brand no longer fulfils their needs. An example of this is the demise of the once ubiquitous photography giant Eastman Kodak. All products follow one or more trends. Some of the trends that are important today were emerging trends a decade ago, but have now started to affect products and services across many categories. A trend may not seem significant to the product category in which the company trades, but it can suddenly affect this category. Documenting trends that do not seem directly related to the category can make it possible to follow and monitor trends that will eventually have a bearing upon the company.

ESTABLISHING SYSTEMS TO MAINTAIN KNOWLEDGE

Past projects and the thinking behind them are important parts of an organisation's assets. Organisations have different systems to capture this knowledge depending on which context they are part of, and depending on the procedures they have developed. Organisational knowledge of particular relevance for brand-building includes: how the customer has interpreted the brand, specific product features, the product heritage and competitive field, past and future trends, and the changes the product portfolio has undergone.

This chapter began with a quote about London Underground. This is a well-recognised brand that has been maintained over decades. As early as 1915 the calligrapher Edward Johnston was asked by what was then called the Underground Electric Railways Company of London to design a new typeface for its signs (Green and Rewse-Davis 1995: 14). He designed the Johnston typeface, which continues to be used on the London Underground almost a century later, with some modifications. Today London Underground is a well-recognised brand, and also part of a major integrated transport system. The Underground, or Tube, consists of many unique features that people recognise as representing the metro system in London. Over time these have become iconic symbols of London itself, and the branding has been developed to encompass all modes of public transport in the city. The core symbols, recognised globally, include the red double-decker bus, the roundel that is used in the logo, and the Underground map.

In 2000 Transport for London (TfL) was created to take responsibility for the British capital's transportation system. Its role is to implement the Mayor of London's transportation strategy and manage the system (TfL 2012). London Underground was included in TfL's remit in 2003. One of TfL's roles is to manage the overall brand of the transportation system in London. It has a small in-house design department that includes graphic, industrial and user-experience designers. Due to its large portfolio of products and graphics, a majority of the design work is outsourced to external companies. To help TfL manage its brand, all its design manuals are freely available online, including a manual covering how to design the different products that make up the transportation system (TfL 2012). This documentation of what a TfL product should be plays an important role in communicating to external design agencies what makes the products part of 'belonging to London'.

An established brand, such as the one Transport for London is managing, may have many design features and qualities associated with it. The consumer can perceive these as belonging to the company. In the TfL case, these features are not perceived as belonging to a particular company, but are part of the identity of London. For tourists, the iconic map, the red colour and the roundel logo 'belong' to London. To earn such recognition, the underlying organisation will have to be coherent in its use of these features and qualities in future products. These can be understood as part of the company's heritage, and also its assets. The qualities and associations customers attribute to the brand that are related to the product must be maintained carefully in future products. Those qualities and features that are associated with the product make up the product heritage.

ESTABLISHING A CULTURE FOR ORGANISATIONAL LEARNING

The design process needs to be informed both by the heritage and the values of the organisation. It also needs to be informed by changes in society. This means that it is important to establish systems within the organisation to promote learning. Larger organisations have often set up 'brand schools' where new employees can learn about the brand. Most have also organised meetings and conferences where employees can come together to offer input and share knowledge that is relevant to projects they are working on. In addition, there is a trend towards consultancies taking an active approach in delivering the information. A common theme for several of the interviewees featured in this book was the importance of engaging clients in active knowledge transfer workshops. Such workshops serve as important meeting points to allow various stakeholders to share information and make decisions. Engaging the client in, for example, the content of the research or defining a direction can serve as an initial step in implementing such workshops.

Members of the business community have proposed 'design thinking' as an alternative approach to addressing business challenges (Martin 2009). This includes a qualitative approach to product development. Establishing practice for design thinking in an organisation may involve methods such as observation of the user or exploration of the context within which the product exists. Qualitative exploration of real-life situations is often adopted as an alternative to the quantitative approach when learning about consumers and the use of the product. This approach may also include developing a sensibility towards the heritage of the product or exploring how the product relates to trends.

Another trend in sharing and exploring practices is the interest in establishing 'communities of practice' (CoP) – 'groups of people who share a concern or a passion for something they do and learn how to do it better as they interact regularly' (Wenger 2006). Stimulating an environment for learning and sharing in the office environment means that knowledge about how the product builds the brand can be maintained and used in the organisation. Both design thinking and CoP need to be implemented organically in the organisation. A first step to implementing these involves studying the core pillars in an organisation. These include the informal and formal knowledge transfer system, decision-makers, motivation for stakeholders to engage, and the organisational structure.

DOCUMENTATION IS IMPORTANT BEYOND LEARNING

Documenting what the product is about – why it was developed as it was, the current market situation and trends – is not only important in order to capture and develop organisational learning, it can also be critical knowledge to capture in preparing for future challenges.

The Norwegian company Stokke found that its children's chair Tripp Trapp was being copied by several rivals. It repeatedly went to court to protect the design and claimed that the chair needed protection as what in law is considered a work of art, not just a patent. In court it was asked to show how the chair represented a significant development when it was launched. Since the chair had been launched in 1972 this proved to be a challenging task for the company. However, it managed to reconstruct the market situation forty years earlier and won the right to continue to enjoy strong legal protection for the design of the chair.

In this case, as in many similar ones, the company won because it managed to provide documentation. This illustrates how important it is for designers to document the design process, the market situation and the trends that they are part of. A good documentation system in a company can serve many purposes. Firstly, it creates a body of reliable reference material to teach future designers about the brand's development. Secondly, it serves as a source of inspiration in future product design processes. It can also make key contributions to the analysis of how the company can protect the design of its products long enough to benefit from return on the investment in design.

From Brand DNA to Design Directions

The design process is one of many activities involved in building a brand. The brand DNA underlies all of these activities. However, in order to make them work in the design process there is a need to translate them into something it is possible to design. How often this translation stage will be needed will depend on the context of the design process. Depending on the goal of the process and the distance between the core of the brand and the individual project, differences will arise. This section will offer a brief examination of some of the reasons for the translation into design directions. These design directions are on a higher level than the individual design process, and will be used to guide design processes over a length of time.

BRAND BUILDING BLOCKS

Brand DNA underpins all activities, including an individual product's design process. The case example of Imsdal bottled water we will examine in Chapter 6 shows a situation where the brand is concentrated on a small number of homogenous products. In this the brand values that are defined may be the same as those that drive the design process. However, brand DNA cannot always be translated directly into design. Companies like Nokia and Volvo Cars therefore devote considerable planning to translating the core brand values into a design philosophy (Karjalainen 2004).

How the organisation is set up and how design forms part of its overall strategy will define the relationship between the brand DNA and the story of the product. The relationship between the brand and product, and how these concepts relate to change, can be described as building blocks, where the lowest layers are less about change, while the top layers are all about change (see Figure 5.1).

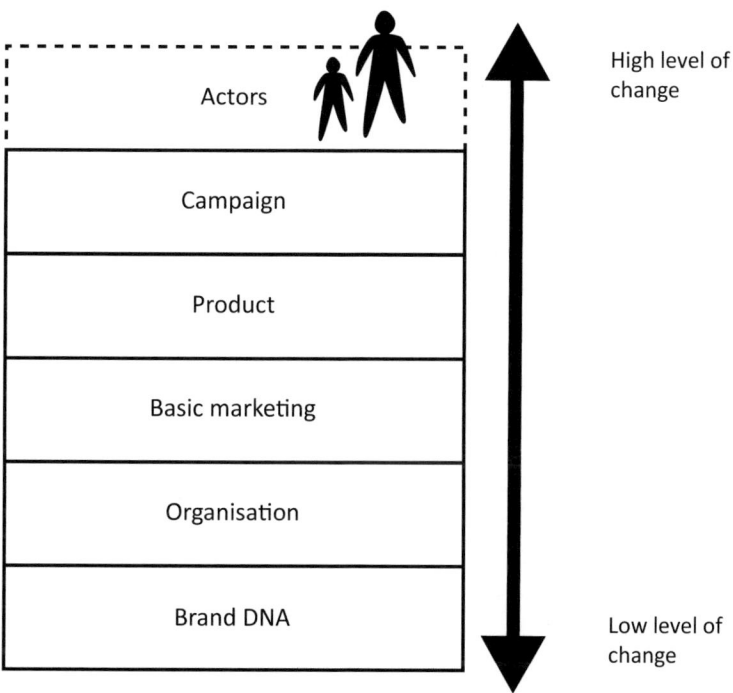

Figure 5.1   The product as part of the brand's building blocks

The brand DNA underpins all activities within the organisation. The values and the visions a team has for a brand define the structure, the information flow, the decision-makers and the motivation within the organisation. Basic marketing involves corporate identity, such as the logo, the typeface or the tone of voice the company uses. The next block is the brand as the product. The following block is the campaign or the promotion activity the company is planning to promote its products. The final building block is the actors outside the organisation, such as journalists or consumers. All of their activities are important in building the brand. From the brand DNA to the actors, there exists a different dynamic. The brand DNA, as discussed in Chapter 4, changes little, while how the actors engage with the product can vary a great deal. The product itself becomes one among many means the company employs to build the brand.

PRODUCT DESIGN AS ONE OF THE BUILDING BLOCKS

In order to build the brand, the company will need to narrow its focus and define the structure which will make what the consumer recognises as the brand (Wheeler 2006). In this the product can play an important role if there is some coherence in how it is designed. The company's design philosophy can be presented as design directions or guidelines.

Design directions will guide each of the design processes. Normally, brand values are defined and translated into design directions. Previous design references in the form can be also be described, as well as examples of what are perceived as the key products within the brand. The design guidelines that inform the design of products should not be the same as the company's corporate identity manual. The dynamics in a product design process are most likely different from those defined in the corporate identity manual. Figure 5.1 shows an example of how the brand building blocks may be set up. It suggests that basic marketing is a less dynamic factor in the process than the product itself. This will depend on the category the product is part of.

In brands where the product is a driving force in building the brand, such as in the fashion industry, the design principles the designer is following may be seen as part of the brand DNA, as these are crucial in understanding what the brand is about. This is also the case in the example of Dieter Rams's 'ten design commandments' we covered in Chapter 2. These ten commandments need to constitute an important part of the brand DNA in order to make as strong an impact as the product did under his leadership. What is important is that the company defines the brand building blocks and the role design plays at different levels. This should include the dynamic of how the product will change with future launches.

DEFINING DESIGN DIRECTIONS

The design directions should be inspirational, and flexible enough to be adapted to a varied range of products. How strict the design guidelines are will depend on the company and how much freedom it allows in interpretation. Guidelines that are too open may not function as guidelines, but more as sources of inspiration for designers. A style guide that is too detailed may stifle creativity at an early stage of project development. If design directions do not make sense to product designers, they will simply ignore them.

London-based design strategist Kevin McCullagh describes how it is important to have empathy with designers and understand what motivates them to engage with the design process:

> *If you do not have enough empathy with product designers, you are not keyed into what product designers find engaging. Often, some of the early concepts that come from a brand or marketing agency are not taken seriously by the product designer. The words they use seem airy-fairy for product designers.*
>
> *(McCullagh 2012)*

The challenge of understanding reports drawn up by marketers or brand agencies also led to frustration for the Norwegian designer Geir Øxseth (2007):

> *I have said it many times: How am I supposed to use these reports? Where does it say anything about what the product should express or look like? When they present mood boards with a million pictures, it looks like a salad. It is hardly possible to draw any conclusions or use the material in your design work. A report that is useful could be three–four pictures, and a few sentences. Preferably some sentences which it is possible to use in a design process. It has to be concrete.*

Directions that are short and concise prove to be more helpful than guidelines that are described in detail or in too many words. Verbosity also leaves scope for many interpretations. Keywords should be defined, together with a suggested interpretation – preferably both verbally and through visual examples.

How often they are updated will depend on how they are described, but also what the brand DNA involves. If a brand wants to be recognised as innovative in the 'look and the feel' of its products, the design guidelines should either be very loosely defined or updated frequently. In the case of traditional brands with long heritages and strong design references that communicate what they are about, the process will most likely be far more dynamic. In this process, the brand DNA (heritage, value and organisation) holds the key to which cultural opportunities may be most appropriate.

DESIGN REFERENCES IN THE PRODUCT

Particular shapes, forms or features of the product can over time be recognised as references to the brand. When customers observe these, they immediately think of the brand. When it comes to defining the explicit design references, it is important to allow flexibility in defining directions for how these should be treated in future products. Design references that are part of the product may include shape, sound, weight or smell. Such references are not as simple to add as the logo. The meaning of these design references must be carefully applied to the designed product in correlation with other parts of it. If the whole of the product changes, the reference will most likely also need to change. The reference will need to be translated into the new form.

Sometimes the reference does not even retain the same form, but it can still embed the same meaning. A curve or a design feature that is defined for one context can create entirely different connotations if placed on a new design. In the design process, these references will have to be brought in and considered. The design references will have to be revisited in the design process and designed in accordance with the product that is about to be developed. It is important that the references have some flexibility, to allow them to be applied to other product categories within the brand. The same meaning will not necessarily be embedded in other products or other categories simply by adding the design references.

Because of the dynamic nature of the brand-building context, the design directions will have to be updated. When customers' perceptions of the product change, the references also need to be updated to remain relevant. If the intended design reference does not follow the same dynamic as the perception of the product, the product will most likely appear old-fashioned and lose its intentional meaning.

ENGAGING THE DESIGN TEAM

In introducing directions and design manuals, there always needs to be a balance between making the directives engaging and inspiring while at the same time providing clear guidance. Inspiring directions will provide leadership, while strict directions will create a sense of forcing designers' hands. Effective leadership is necessary to develop a culture within the organisation that attracts the desired talent. In 2010 what was then the Swedish-Japanese mobile phone manufacturer Sony Ericsson (from 2012 wholly owned by the Sony Corporation) wanted to draw up new design directions for its products. From a designer point of view it managed to find an effective way to engage the whole organisation globally with these new directions (Movold 2011). Firstly, it developed a theme and values based on cultural insight and analysis. In the next stage it introduced a design competition where all the designers worked in teams of two towards a deadline to develop concepts according to the theme. The whole experience was designed as a competition. At the end of the process the different concepts were presented to the top management and the designers were given the opportunity to present their visions and interpretations of the theme. After assessing the different concepts based on design quality, innovation, marketability and feasibility, a winner was chosen to become the starting point for developing the next year's form language.

In the end a team from Tokyo won the competition, but everyone had received an opportunity to both pitch and work creatively at the start of the design project. Cathrine Movold, a former planner for Sony Ericsson who shared this story, felt:

> it was OK to work on the execution of this form language for the entire portfolio [over] the next ten months. It had been an open process that was inspiring to work on, and the team in Tokyo had an excellent concept. This is different from having a framework that is forced upon you.
>
> (Movold 2011)

By engaging a larger number of members of the organisation in a competition, Sony Ericsson managed to encourage and inspire the designers to work in the same direction. The approach also made it possible to involve all stakeholders who were part of the design process while at the same time avoiding the need to make compromises that might diminish the vision.

Strategies for Developing a Strong Brand Identity

How a company uses design will be linked to its choice of designers and how its design work is organised. The designers who translate the brand DNA into tangible products will have an influence on both the idea behind the product and its look and feel. Selecting the right designers or establishing the right design team will therefore be major decisions for the company, and are important when defining how design is to be used as part of the overarching brand strategy.

The choice of designer for a specific company is governed by the context within which that company operates. It also depends on what it wants to achieve when developing a product, and what resources are available. One major decision is whether the company should have an internal design department, or work with one agency, or with a pool of designers. Where a product has a strong brand identity, it is important to establish a strong design culture within the company.

TRAINING IN-HOUSE DESIGNERS

The Dutch designer Guido Stompff (2008) recognised the importance of having an in-house design department in order to develop a brand through the products. There are many benefits from building a brand through internal design resources. Branding and design rely on knowledge. From this perspective, an in-house department can build a culture around the design of the products. The company will have better control of knowledge about the brand, and will also reduce the risk some companies face of becoming too dependent on individual external designers (Jevnaker 1995). Without knowledge about how to lead designers who have a strong design philosophy, companies can be vulnerable. Not all designers take a strategic approach to the design process, and the philosophy of what the product can represent may be more important to them than consideration of what will be beneficial in building the company's brand. With an in-house design department the company will have greater freedom for trial and error when experimenting with the role the product can play in building the brand before launching products on the market.

A major benefit of having an in-house design team is that by immersing themselves in the company culture over time, the designers become part of that culture. This means that the decisions they make when designing are manifested in the brand. In the case of Nokia, it took as much as two years to train designers fully about the look and feel of the brand (Karjalainen 2004). By being part of the company culture, designers can gain a clearer sense of what the brand is about. These are lenses they can use when adapting to shifting circumstances.

In building a design team, it is important to recruit people who share, or at least understands, some of the same values in order to develop a coherent value set. In situations where the company seeks to change direction recruiting people with new value sets can help move the brand in the desired direction.

Many organisations choose to have an in-house design department while outsourcing some aspects of the design process. The organisation may not have access to specialised services it needs in-house, or it may simply want an outside perspective to question what it is doing internally. All organisations will to some degree find that they have established 'truths' which do not always correspond to the real world. External input can thus be valuable in order to adjust or fine-tune the direction of the design work.

ESTABLISHING LONG-TERM RELATIONSHIPS

Ringnes, a subsidiary of the Carlsberg Group that produces soft drinks, beer and bottled water, sees itself as a market-driven company (Horsrud 2007). The design of its bottles plays an important role in its strategy, but designing products is not a core function of the company. To maintain focus on its main role, it outsources its design work to design agencies that help it to stay in touch with current trends and offer fresh views on product design. To balance this with the need for a coherent look and feel over time, Ringnes builds up long-term relationships with its external consultants. These design companies work on the same brands over extended periods, which means that they are familiar with the brand DNA and product DNA. The design agencies also learn about the aesthetic presence of each of the brands they are working on. Each agency almost becomes part of the company, but the relationship also tends to be based very much on personal connections. A group of marketers or a single marketer from within the company will often collaborate with one or two representatives of a design agency, developing relationships over time.

USING A DESIGNER'S SIGNATURE AS PART OF THE STRATEGY

In product design, the various design schools have placed less emphasis on designers' signatures as the profession has been influenced by ideas about what the 'right' design approach might be, as we discussed earlier. However, the different schools, traditions and signatures will form a foundation for how product designers approach their projects. This defines who they are and what they emphasise in the product design phase. For this reason, giving the same brief to a range of designers will result in a range of interpretations of it.

How design agencies deal with these different backgrounds varies. Some design agencies have decided to narrow their scope to rely on only one approach, which may become their signature. As we saw in Chapter 4, Ergonomidesign has adopted a focus on ergonomics as its core approach, which is highly visible in the products it designs. Other companies have more variety in what they offer, adapting to the client's DNA. Different design agencies have different agendas, and their individual signatures are important elements of the design assets and skills they offer to their clients.

DISRUPTION FROM AN EXTERNAL AGENCY

The new bus for London (NB4L) is a double-decker which draws its inspiration from the original classic Routemaster 'London bus' built by the Associated Equipment Company (AEC) from 1954 onward and in ordinary service until 2005. Working prototypes of the new bus were launched to the public in December 2011. This was the first new bus that had been designed for the capital in more than fifty years (TfL 2012), and has been a prestigious project for the Mayor of London, Boris Johnson.

The project began when *Autocar* magazine commissioned the English transportation design company Capoco to develop a 'new Routemaster' (*Autocar* 2007). *Autocar*'s competition was followed up by Boris Johnson with a public design competition that was announced in his Election Manifesto of 2008: 'London needs a fresh perspective. I want to introduce a 21st century Routemaster that will once again give London an iconic bus that Londoners can be proud of' (Johnson 2008). The competition had two finalists: a bus from Capoco Design, and a collaborative design from luxury car maker Aston Martin and the world-renowned architectural practice Foster + Partners.

In January 2010 the London-based design firm Heatherwick Studio joined the design team in charge of this project. They were tasked with developing a new London bus that would eventually go into production by the Northern Irish company Wrightbus. This also meant that all aspects of the finalist projects from the public competition were not taken forward. Paul Marchant, Head of Product Design at Transport for London, stated:

> *the brief for Heatherwick Studio, who are an external agency, was to design a new bus for London and not necessarily take someone else's concept forward into the detailed design phase. Because of the project's ambition Heatherwick were encouraged to place their own signature upon the design of the bus.*
>
> *(Marchant 2011)*

Heatherwick Studio has a strong design signature, and its projects have a similarly strong character. This makes it difficult to unify its designs with earlier approaches, such as those resulting from the design competition. Heatherwick Studio is known for 'visionary' design that attracts a wide audience. Some of its best-known projects are the London 2012 Olympic Cauldron and the UK Pavilion (the 'Seed Cathedral') at Expo 2010 in Shanghai. Appointing a design studio like Heatherwick means that there is a desire to develop an installation piece of work. The prototype of the new bus for London that was launched has these qualities, and it certainly has a different look and feel from other products in Transport for London's portfolio. It can be challenging to incorporate projects like this into an existing holistic design framework like TfL's.

Transport for London has a small in-house design department and outsources the majority of its design work. For an organisation such as TfL, its relationship with external agencies tends to be long-term. In considering a new agency, it looks for those with at least seven to ten years' experience in transportation design (Marchant 2011). TfL recognises that this maturity in its field of design also brings an understanding of the need to comply with the standards that regulate the transport industry. Even with this background it takes time to build up the necessary knowledge of what makes a product right for London. Paul Marchant of TfL said: 'Whenever you start with a new design agency, it takes at least two or three years before they understand what we are and what we are doing' (Marchant 2011).

The internal design team at TfL maintains a knowledge base about the different design agencies. Part of the relationship involves understanding the agencies' weaknesses in order to decide which design agency will be appropriate for a specific project. The in-house design team plays a crucial role in managing the brand, and therefore participates in decisions about every new product that is designed.

TfL's design team was not directly involved in the process of designing of the NB4L prototype. The look and the feel of this bus are quite different from other products in the TfL portfolio, and some of the design details present challenges. Cost, functionality and aesthetics are carefully balanced in transportation design. The overall look of a London bus is due to the demands the bus will have to meet. If you make aesthetic decisions in isolation without balancing them against costs and functionality, you may encounter problems. Marchant expanded:

> If you look at the back of the NB4L there's a large glass sweep which is quite elegant. However, if someone bumps into the back of it, that glass is going to break relatively easily. It's a big piece of glass, so is going to cost and take time to replace, and then that bus is out of service.
>
> (Marchant 2011)

The NB4L project is inspirational and visionary, but the design has received a mixed reception from the city's inhabitants. This is to be expected when choosing a design that is regarded as a disruption of the traditional look and feel of transportation in London. The widely read *Londonist* blog observed: 'flowing forms of the exterior are matched inside, and every detail, from the recessed lights to the "stop" buttons, are clearly the product of a heightened aesthetic sense' (The Londonist 2011). Jonathan Glancey, *The Guardian*'s Architecture and Design Correspondent, was less appreciative, stating that there was 'something not quite right about it' and that the bus had 'too many swirls' (Glancey 2010).

A bus with such a radical look and feel could lead to fragmentation if not carefully managed and followed up, particularly in an organisation with such a well-developed design framework as TfL. This is why the follow-up work on the NB4L will involve a more important role for the in-house design department. The NB4L is an example of a project that creates enthusiasm, and could be seen as reflecting the Mayor's vision for London as a city. The bus incorporates references to the past, but the form breaks with this heritage far more than the other solutions that were presented in the competition.

WORKING WITH DIFFERENT AGENCIES AS PART OF THE STRATEGY

Enlisting different designers as a strategy to build competitive advantage has proven to be a successful approach for companies like Alessi (Verganti 2010). The Norwegian oral hygiene company Jordan has also used this to its competitive advantage. All the staff interviewed at Jordan emphasised that their company is market-driven, not design-driven, although the design of the product plays a major role in the brand-building strategy (Hestad 2008). Jordan is competing in an FMCG category that is dominated by a few global brands owned by major multinationals: Colgate (Colgate-Palmolive) and Oral-B (Procter & Gamble). These companies have a much larger marketing and research budget than Jordan, a relatively small company owned by the same family since 1837 and only taken over by the branded goods conglomerate Orkla in 2012. Faced with this competition, Jordan has chosen to use 'design' as its competitive advantage. What 'design' means for Jordan is not defined in terms of strict design guidelines, but it has an overall idea of what it is. For Jordan, it is associated with 'Scandinavian design', and an interpretation of this that leads to simplicity in its toothbrushes. This also means that it has a very clear idea about how its toothbrushes should *not* look.

A core part of Jordan's strategy is to use different designers for each toothbrush. Using different designers and letting these designers develop the form based on their own styles has created products with a broad spectrum of personalities, as the designers have different signatures. Jordan starts with a clearly defined market niche or a cultural opportunity in the market. It then selects designers according to who it thinks is most likely to design a toothbrush that will meet these challenges.

A clear understanding internally of the role of product design in building the brand is the reason why working with different design signatures has been a successful strategy for Jordan. The process is led by the heads of the Marketing and Product Departments, who work closely with outside designers and guide them through the process. Both have been instrumental in defining Jordan toothbrushes' look and feel. The role of these members of staff is very important in maintaining the company's focus on the product's role in building the brand. Without someone inside the company who has this focus, it is challenging to maintain coherence in what the brand is about.

In the case of Jordan, the two people in charge of the in-house team and liaising with the design agencies did not come from a design background. However, they had an appreciation for design and a clear idea of what designs would fit in with the Jordan brand. They serve as guardians of the look and the feel and how the idea of the product corresponds to the brand DNA. The role of the guardian is not to conserve the past, but to maintain the brand so that products remain relevant to consumers.

A Culture for Design

Harnessing the design process as an important factor in building the brand cannot be accomplished overnight, and is an iterative process. Past products and past activities become the heritage of the brand, and this heritage should be treated as an asset for the company. Each of the design processes in branding represents a case for learning about the brand. By establishing a culture for organisational learning, the company could make use of this knowledge in future design processes.

The brand DNA needs to be translated into design directions and guidelines. How these are shaped will depend on the context of the design process. These directions will function at a strategic level, providing guidance for the various individual design processes. However directions will not be enough to maintain and build the brand. More important is the culture in the company. The selection of the design team is an important strategic decision and the values the team share will create the premise for the final design. If some of the design work is outsourced, there is still a need for people in-house who have an appreciation for design and understand how it relates to the brand DNA. In order to develop products that represent the brand, the company will need to nurture their culture.

Research: Informing the Design Process

We see that companies are missing out on the cultural opportunities within the product and the brand. By conducting research we are able to tell why people are behaving in a particular way – and embedding that knowledge in the design process just gives us a bigger context. An example is reuse of clothes. People are taking their children's clothing and the clothes are handed down to another family. In order to get a deeper understanding why this behaviour is so common, and why people feel there is so much value in it – you have to look at the context of parenting and the value attached to behaviours embedded within a particular community. Then you start to understand why this is happening. Such insights then can lead to new design outcomes. This approach shows that there is an opportunity for re-thinking the design process.

Joanna Brassett, Founder of INTO [Studio]

Information is of key importance in the professional design process. This chapter will look into what kind of information the company has access to, and how this could inform the design process. The use of the product will also contribute to the perception of the brand.

Information from the Market or Society

The design process includes a series of key stages, from developing the first design brief to industrialisation of the product. The design and innovation consultancy IDEO has simplified these stages into three: inspiration, ideation and incubation (Brown 2008). Inspiration is the phase that defines the direction. The ideation stage is where various concept ideas are developed. The final stage, implementation, involves making the chosen concept real. Major decisions at all these stages will affect how the product builds the brand.

Research is an important part of the design process (Cooper and Press 2003). It is often used as a source of inspiration, both to identify new opportunities and to go deeper into a desired direction. Research could also be used to validate ideas. An important consideration is whether research should include user research, and if user research is employed, at which phase in the product design process it should be carried out. Finally, it is important to consider how the information from the research can be used to inform the process (see Table 6.1). Research, both internal and external, is an important factor in making informed decisions.

Table 6.1   Example of the deployment of user research at different stages in the design process

Phase:	Inspiration	Ideation	Implementation
Types of research	Consumer research Technology or material research Socio-cultural research Research into theme – could be historical, philosophical Research into own brand heritage and DNA	In-depth user research Research into material, technology, colours, trends, competitors Research into production techniques	Design testing Research into production techniques continues
Aims	Establishing direction in the project	Better understanding of the use of products that have been developed Detailed research into materials and technology to inform decision-making	Testing whether incremental changes are in accordance with perceived product DNA

A majority of the research is carried out at the beginning of the design process, before concepts are developed. In the first stage of the design process, research can define opportunities and provide information about the company's understanding of the brand. Later, its function will be to arrive at a better understanding of the process and what consumers think about the products that are designed. These processes will overlap, and designers will find that they need to conduct additional research in the ideation stage, and even in the implementation stage. In the implementation stage, companies also conduct design tests, to check whether the approaches chosen are appreciated by consumers.

The increased use of research in marketing and design is being questioned in both sectors. There is a debate about whether the use of market research has led to an increase in 'me too' products, where the product that is developed is purely a response to what is already on the market (Zaltman 2003; Beverland 2009). In the design community there is also an ongoing debate about whether user research ever leads to innovative products (Norman 2009). The argument is that studying already existing products and how consumers are using them tends to result in improved products, rather than new ones.

TECHNOLOGY CREATES OPPORTUNITIES FOR NEW USER SCENARIOS

The American academic Don Norman argues that technology comes first, products second, and needs third: first, the technology is invented, then technological inventions drive the development of new products, and when these products finally come into being, the need for them becomes apparent. It is in the final stage that design research can play a major role, according to Norman. In his article 'Technology first, needs last: The research–product gulf' (2009) he lists products such as the radio, the phone, the car and similar ones that all follow the 'technology first, needs last' model. Many of the products he refers to were invented long before design education was formalised. This explains the absence of a design research basis for the products he mentions, but it does not make an argument against basing other products on design research.

Bruce Nussbaum, an editor of *BusinessWeek*, responded to this debate by calling Norman's view of innovation 'top-down, one-way and very old' (Nussbaum 2009). Nussbaum argued for social innovation, and pointed out that inventions need to have a social dimension in order to become innovations. Looking at products that are regarded as innovative, it is evident that technology- or material-driven inventions need a user scenario in order to become innovations. This is an important discussion, as it points to *how* research is brought into, and informs, the design process. Norman has an important point, in that a designer who employs user research directly in the design process will often come up with an incremental change in a new product compared with an older one. However, as Nussbaum argues, designers need to understand the social context the product is part of. Without this, the product may be a fatal flop or not even be developed in the first place.

THE IMPORTANCE OF USER RESEARCH

Norman (2009) highlights the importance of user research even though it tends to lead to incremental changes. In brand-building, these incremental changes are as important as major ones. Every time I switch on my television set I am reminded of the importance of user research. To switch on my television I need two remote controls. Navigating the menu to find the right channels is almost a hopeless task – not to mention the slow and difficult navigation of the 'programmes on demand' function. This is part of my everyday experience with my TV, and it has definitely affected my attitude towards the cable TV company that delivered the set-top box. The promise the company behind the brand made was 'fun and innovative'. My associations with the brand now are 'hopeless, gimmicky and technical'. The service and the product that provides it do not deliver the expected experience. Luckily for the company, most consumers are probably more forgiving than I am when using the TV. According to the researcher Don Norman, users tend to blame themselves rather than the product (Norman 1998: 37). Nevertheless, by not delivering the brand promise through the experience of the product, the company is opening up an opportunity for others to deliver that 'fun and innovative' experience.

In branding, as described in Chapter 3, the whole user experience needs to be taken into consideration. Over several years the British consultancy Plan Strategic has built up expertise on how consumers experience products. These experiences will not only result from the tangible object, but a combination of all the touch-points consumers have with the brand – including their own use of the product (McCullagh 2012). Companies that want to arrive at a comprehensive understanding of how consumers perceive their products and brands can conduct consumer experience research. This could focus on what Chris Rockwell (2010) of the American consultancy Lextant calls the 'consumer experience journey' – mapping the process step by step. Some of the steps along this journey are more important than others. A key stage is before consumers have made their decision to buy a product. Other important stages occur at the point of sale, and when customers actually use the product. This step-by-step process helps companies to understand how consumers experience the product over the course of the journey. It can also be used to analyse similar processes that competitors face. By analysing the product experience step by step, the company gains an insight into how well the brand is performing and how well its competitors are performing. An experience curve can be drawn based on analysis of the intensity products excite on this step-by-step journey.

Through this approach, the design team can also identify which stages in the customer's journey with the product provide the highest value, so it can focus on the most important ones in building the brand. This may involve key signatures or other design references that can be embedded into the product. These factors can be followed up, either by advertisements or by design references in the product itself. By co-ordinating the marketing and product in key references, over time these may come to be associated with the brand. The brand will establish ownership of these. However, companies should be careful not to choose design references that feel artificially added or forced, as these may be less well received.

An important part of understanding how a product helps to build the brand will therefore be to explore whether there are any brand gaps, such as between how consumers understand the product and what the company intended the product to look like. The gap may also be internal. There may be disagreements in how internal departments understand the brand, and there may be a difference between what the marketing material says the brand is about and how the design department defines it.

When conducting research into how consumers perceive products and brands it is also important to cover what the company has already successfully built. Some products will not be based on changes, but simply to maintain what has already been established. The dilemma all companies face in finding the right balance between coherence and change will be revisited in Chapter 7.

RESEARCH TO VERIFY THE DIRECTION

This chapter started off with a quote from the founder of INTO (studio), Joanna Brassett. After years of working with global brands, she feels that companies generally have a good understanding of markets and trend research, but they often miss out on opportunities in the cultural context. She believes that companies need to focus even more on embedding cultural understanding into the design process. By exploring how people behave and exploring these socially embedded behaviours in greater detail, a company can define a direction for the design process. This approach differs from traditional user research. In traditional user research the focus is on how people are using a product. It is possible to develop a better user interface by exploring how a user interacts with a mobile phone, for instance. Looking at the cultural context can include investigating why people fiddle with their phones while waiting for someone in a café, or why people often keep their old telephone when they buy a new one. These types of insights into how people behave can provide important starting points for the development of new user scenarios.

Before starting a new design process, the design team needs to have a clear understanding of the drivers of change and trends related to the product and the brand. This is important for many reasons. It is vital in order to help the company to stay ahead and remain relevant to customers. Furthermore, this input can be a tremendous source of inspiration when developing a new concept. The research can help to validate the product that is designed.

It is important to establish a sense of direction when designing a new product. This will depend on what kind of product is being designed and how it is intended to build the brand. The impetus may result from an opportunity that has been identified in the market, as in Jordan's efforts to design an ergonomic toothbrush or a toothbrush on-the-go described in Chapter 4. The direction may also be defined by technology and new user scenarios new technologies give rise to. It may result from a new trend, or a cultural opportunity that the design team has identified. There are various approaches to defining this direction. However as Jordan found, the company will have to adapt to unexpected user scenarios that may appear when it begins to follow the direction.

Positioning is an important stage when building a brand. It will also be an important factor in defining the direction of the brand. The American design agency Ziba sees the visual positioning of products as an important aspect of developing a distinct portfolio with brand recognition (Vossoughi 2007). It does this by mapping its own products and competitors according to relevant keywords that drive the look and feel of the products. This exercise can be useful to gain a better understanding of the company's own product DNA in comparison with other products. From time to time it can also serve as an eye-opener if the product is already competing in a market with what appear to be 'me too' products and is not adding any new value.

Visual mapping can be an important tool in enabling products to succeed, as it allows the design team to establish what else is out there, and whether the product is different from other products in the current marketplace. However, if this information is used to define the direction, there is a risk of missing opportunities outside the market space. If a company wants to establish itself as a distinctive brand, it needs to position itself carefully in the market. Starting from this perspective means that the market will define the premises for whatever the company is seeking to build. On the other hand, once the direction has been defined, positioning can be used to further define the brand, clarifying it and defining future positions that are desirable. This is an important part of the strategic toolkit when working on defining how product design can contribute to building a brand.

In organisations where large teams are involved in the design process there is a particular need to establish common ground and a clear direction. It is important to establish a design direction as a common vision of what the product should be, but also how the product can help to build the brand. The direction should be the meeting point for all stakeholders involved in the process, and guide the design team in all its decisions. Research can help to verify the direction and provide a rationale for choosing it. The direction should be an aspiration for the team members, and is important in providing encouragement to the team.

TREND RESEARCH

Knowledge about trends is essential, not only in order to maintain the brand, but sometimes in establishing whether the key product is relevant to the current (and future) market. A major change in legislation or government policies, in technology, in the economy or similar areas may change people's behaviour. An example of this can be found in music consumption. Apple introduced the iTunes Store in 2003, two years after the launch of the iPod. This combination of a product and a service changed how people bought music. A similar driver of change is the simplicity technology has created in downloading and buying digital books (for example, Amazon's Kindle). So far, this has not overtaken people's desire to buy hard-copy books, but the ripple effect seems to be spreading, with more and more companies focusing on digital books. The last three months of 2010 saw the sale of Kindle ebooks surpass paperbacks for the first time (Barnett 2011).

Brands that are market leaders have often been established during a major shift in the market. In many cases they have been at the forefront of innovation during the shift. This provides these brands with a unique position that can be hard for other brands to follow, since they will always be followers of the market leader. One strategy to overcome this is to look for new potential shifts and create new categories. For certain brands it may be better to follow trends that have already been established by other brands or categories. There can be a thin line between following a trend and producing what will be a 'me too' product.

Understanding underlying changes in society rather than just following visual trends gives designers an opportunity to come up with solutions to design challenges that are in tune with, or at the forefront of, what is happening in society. Looking at trends that are already happening within the same category will only make your brand a follower, not one that leads the field.

CHALLENGES WHEN TESTING DESIGN CONCEPTS

A design test is a form of user research that seeks to establish whether or not users like a product. They are not conducted to learn about the users themselves, merely to establish whether they find the design acceptable. Design tests have been criticised for being of little assistance, particularly when developing radical innovations (Zaltman 2003). There is much evidence to suggest that design test results are not always borne out when the product is launched on the market. Michelle Wentworth, Category Development Manager at Jordan Dental Products, cited the example of the choice of colours for its toothbrush. Market research had suggested that customers would prefer the new toothbrush to be black and white. However, when the brush was released on the market, the sales were mainly of green and pink models (Wentworth 2007).

The test in question simply set out to investigate what would be the right colour for the new toothbrush. When it comes to more challenging questions, the tests can be even more unreliable. Nevertheless, they are important for companies. The board of directors will need facts and figures on which to base its decisions, and if implemented correctly, market tests can provide some indications of the best direction for the company.

When it comes to well-established brands with strong design references, design tests can be more important. In a design process geared to giving a product a facelift, it will be important to find out how devoted consumers are to the existing product. A company may have successfully built a key feature into one of its products that is recognised by consumers and plays an important role in the story-telling. In such a case the company must recognise that feature as a valuable asset, rather than introducing change for the sake of change. As soon as the company decides to change a key feature, it provides an opening for its competitors to adopt it, as it is no longer distinctive. The importance of maintaining what has already been built will be covered in more detail in Chapter 7.

Product Design as a Response to Change

RESPONDING TO CHANGES IN THE MARKET AND SOCIETY

Chapter 4 presented a model of how the design process could fit into the larger framework. In this model the design process depended on information from the company (brand DNA translated into design directions), from the market and from society. What kind of information the company is using – but more importantly, how it applies this information in making decisions – will be important in shaping the brand. A market leader will respond differently than a company that is a follower, and needs to stay ahead of changes in the market. However, market leaders will find themselves in challenging situations because others with less strong positions will seek to seize any opportunity that changes consumer behaviours. This challenge can be exemplified with a case from the bottled water industry.

Imsdal is a Norwegian bottled water brand produced by Ringnes, a subsidiary of the Carlsberg Group (Hestad 2007; Hestad 2008). Imsdal was the first bottled water to enter the Norwegian market successfully, thereby gaining the position of market leader. In 2000 the company experienced a challenge when the market for bottled water changed. Its main competitor, Bon Aqua (a Coca-Cola brand), had launched a new bottle with a design that immediately appealed to young people.

The team behind Imsdal knew from market testing that consumer preferences for bottled water had changed (Horsrud 2007). While consumers used to favour Imsdal because of its associations with purity and indigenously sourced water, the bottle was seen as 'dull' and 'traditional'. An important factor in this change was a trend in society labelled 'body beautiful' by the Scandinavian Design Group agency (Gran 2007). This meant that consumers were aware of the importance of a healthy lifestyle – not as a diet, but as a drive to lead a better, more balanced life. Drinking bottled water was one way for consumers to signal to their peers that they had a healthy lifestyle and took care of themselves. Bottled water had therefore become a fashion object – and the Imsdal water bottle was seen as too traditional.

Figure 6.1 Imsdal, 2002   Figure 6.2 Imsdal, 2006   Figure 6.3 Imsdal, 2011

Photo: Halvor Grønli.

Ringnes had to adapt the Imsdal bottle design to these new preferences in the market. However, being the market leader, it could not change it in response to its competition (Horsrud 2007; Gran 2007), it had to come up with its own interpretation of the trend. The first step was to look deeper into current trends and the values the Imsdal brand represented. The company decided to reposition these values. Imsdal's original core values were defined as 'Norwegian', 'pure' and 'natural'. In repositioning the values, it strived to communicate 'Norwegian', 'contemporary' and 'purity', retaining 'purity' and 'Norwegian' as core values, but seeking to shift from being seen as traditional and dull to being perceived as contemporary. By taking a step back in the process and looking at current trends through the lenses of its own values, Imsdal managed to design a new bottle, not in response to the premises of the market, but to cultural changes (see Figure 6.4).

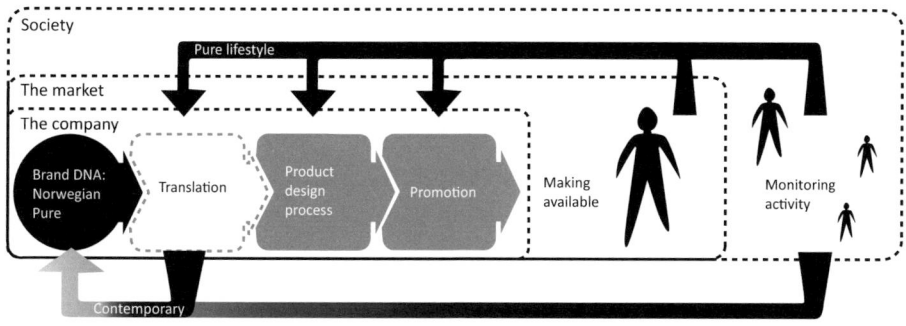

Figure 6.4   **The Imsdal design process – Brand Activity Framework (author's reconstruction of Ringnes's process)**

RESPONDING TO CHANGES WHILE MAINTAINING THE BRAND

Before the redesign, Imsdal had identified five design references customers recognised as part of the Imsdal brand: a mountain-shaped form in the base of the product, the blue colour, the Norwegian flag, the shape of the label and the logo. An important point for discussion was whether it should change any of these references, in particular the 'mountain' in the base. During the research for this book, one response to this element was:

> *Every time I leave Norway I buy an Imsdal bottle, and when I am on the plane I look at the bottle and observe the little mountain in the bottom of the bottle. This is part of my goodbye to Norway.*

This personal and emotional story from an Imsdal customer explained the significance of the mountain. However, in the redesign process the company decided to move away from this design reference, despite market research and advice from its external partners (Hestad 2008). When Imsdal was re-launched with the mountain on the label instead of the bottle, it lost ownership of the symbol. This allowed its competitors to launch bottles with a little mountain in the base, meaning that Imsdal lost its unique brand identifier.

The new references that were identified and the value concepts these represent are presented in Table 6.2.

Table 6.2   Imsdal design references in the 2006 bottle

Element	Description	Value concept
	Imsdal	Pure, Norwegian, contemporary
	Imsdal logo	Imsdal
	Reflection, logo	Pure, contemporary
	Norwegian flag	Norwegian
	Mountain	Norwegian, pure
	Tall silhouette	Proud, contemporary
	Blue	Pure

In 2011 Imsdal re-launched its water in another bottle (Figure 6.3). The changes in the 2011 bottle can be seen as incremental ones that follow the same value propositions as the 2006 bottle, except the shape has slightly sharper shoulders and the label has been straightened. In this bottle Imsdal reintroduced the 'mountain'.

THE DISTINCTION BETWEEN MARKET AND USER NEEDS

Most branded organisations monitor consumer activities closely. Each of the brands researched when writing this book engaged a brand agency or a professional marketing agency that was responsible for monitoring the brand. Depending on how the brand has evolved, it can usually be monitored on a day-to-day basis. Charlene Li and Josh Bernoff at Forrester Research described this in their book *Groundswell* (2008). They show how a system can be developed to enable companies to listen to how consumers are talking about their brands on the Internet, as well as organising more traditional surveys to establish consumer sentiment. Consumer research on its own is not enough to inform the design process. Such reports may tell a story about a brand that is relevant for communication, but do not necessarily provide relevant and sufficient material for the next product the design team is about to embark upon.

The designers interviewed pointed to market research and segmentations as being useful in establishing a starting point for the process, but felt that that these tools were inadequate when designing new products. The customer profiles and target segmentation presented in marketing reports are too shallow to promote deeper understanding. Cathrine Movold (2011) explains:

> *it is important to talk with real people, and understand how they are living. In order to design something that is good, you have to have empathy with the user. You don't get that from reading a report stating that the target segment has 2.1 children. You get it from talking to the people you are designing for and trying to understand their needs and wants.*

As a product planner at Sony Ericsson, Movold also experienced the difference between the user perspective and the market for mobile phones. According to her, not everything that happens in the market is a response to user needs:

> *[In the] technology sector there was an extreme feature race before the iPhone was launched [in 2007]. The primary focus was on the camera.*

It is of course useful to have a camera on the phone. However, at one point some feature phones had cameras that took so high resolution pictures that the user could only save a few pictures to the phone's internal memory. That was not practical nor user friendly, but the focus was only on specification, not on user-experience, and this was a development driven by the constant response to the competitors' specifications.

(Movold 2011)

In her experience there were internal truths in the company that were strengthened by changes in the marketplace. One such truth was that 'people want high-resolution cameras'. These were not related to real insights about how people perceived their phones, nor what they wanted from them.

Building a brand is a cultural process. People engage in branding because it gives them meaning (McCracken 2005). In order to design products that will build the brand, it is crucial to have empathy with potential users and develop an understanding of the cultural context the product will be part of. Market research and consumer research are important aids to the design process. It is important for every company to keep an eye on what its competition is doing and understand how consumers perceive the brand. However, the key issue is how it makes use of this research. If market and consumer research are used at the start of a design project, the new product will most likely be a response to what is already out there. If the direction is set first, then using the research to establish a position or to enable even stronger differentiation is helpful in order to keep up-to-date to what else is out there. The use of market research will also depend on the category within which the product operates and the company's traditions regarding such research.

Informing the Design Process

BRAND AWARENESS IN ALL DECISIONS

Chapter 3 set out a list of strong product stories and key criteria for defining product DNA. Each and every product that is designed needs to have a direction of its own that is relevant to the current context. At the beginning of the design process it is important to ask why there is a need for a new product, and to picture how it can play its part in building the brand.

The design team needs to develop a strong direction that drives the design process. This search for the 'big idea' behind the product is an important aspect of every design project. When seeing the product design process as part of building the brand, it is important to ensure that this direction is aligned with the company's intentions for the brand. If the brand is about ergonomic products, the direction also needs to encompass the ergonomic. If the brand is responding to contemporary art, the direction needs to encompass the philosophy it responds to. In so-called 'product-led brands' this is a natural part of the product DNA. As in the case of Imsdal, the direction may also be a response to a trend that is relevant to the target audience, but through the prism of the brand DNA.

The design process often begins with identifying an opportunity, either in a cultural context or within the market. This can be the starting point for defining the direction. However, the process will need to be flexible. This is particularly evident when translating the direction into a physical object, which can run the risk of failing to communicate what the brand is about. In the Jordan case presented in Chapter 4 there was an example of how the design company translated 'ergonomic' into a toothbrush that did not have the look and feel of the brand. Although the opportunity identified in the market was right for the company, how this was translated into a material object was not right for the brand. The toothbrush's appearance became too aggressive.

An important consideration in the design process is to incorporate awareness of the brand into every major and minor decision that is made. Major decisions in the design process may define the opportunity and the direction, and select the concept and any variant of a form. Each of these decisions will have an impact on how the product helps to build the brand. The design process is iterative, which means that the designer will go back and forth, and learns from each stage in the process. However, the major decisions are regulated by constraints such as time and the resources available. When a major decision is made, the team must take into consideration how the product it develops will build the brand. This vision is of key importance in building a solid argument for whether or not the company should go forward with a particular product. Developing a product that is aligned with the idea of the brand is not always the cheapest solution for a company. However, the long-term gain from developing a brand rather than just producing a product will be worth the investment.

When the direction has been set and the design team has a coherent understanding of how the product can build the brand, brand awareness ideally needs to inform every minor decision. The product's main form, its silhouette, its finish and detailing are all important aspects of the product, and the latter two are essential in giving a product the right look and feel. For high-quality brands, the attention the design team has paid to detailing and the finish of the product is what communicates the level of sophistication that is expected. The shade of a colour or the texture of a material are perhaps not perceived as major decisions, but paying attention to all these details will make a great difference to the product. It will convey an impression of being carefully designed. The complexity of decision-making in the design process is why it is so important for the process not only to define the brand through the product, but to establish a strong emotional connection between the designers and an ideal of how consumers will experience the new product.

Developing this brand awareness is not so much a question of considering every minor decision very carefully, but of building awareness and a strong vision of what the product should be. Establishing a strong image of how the branded product should appear early in the process is one way to work towards building brand awareness.

In the Imsdal case the team developed a visual presentation of the core values the company had defined and passed this on to its industrial designers. The company first defined the opportunity it had targeted and developed a clear understanding of why this was the right opportunity for the brand. Later, together with designers and marketers, it developed a simple presentation of the brand's core values, defined by three images and by three key values. The clarity of these was also recognised by the designers who translated the values into the package. The key principle in developing visualisations of a brand is that this needs to be carried out in co-operation with both designers and product and brand strategists. When all parties are satisfied with what has been defined, the project will have a clear visual direction.

KNOWING THE CONSTRAINTS OF THE PRODUCT'S CONTEXT

Understanding the product's contextual constraints is vital in harnessing the product design process to build a brand. There are constraints in every company, in the market context and in how people will use a product. In the Imsdal case there were many constraints in the product life cycle, from designing the bottle to using it, that had implications for the bottle's design.

For example, in the manufacturing process there will be constraints on making bottle from a certain type of plastic. On the drawing board the bottle may have a dynamic that is not always possible to implement because of all the standards the bottle will need to meet (Horsrud 2007). The bottle will need to follow production standards when being filled. For Imsdal, this meant that the bottle would have to have a ring around its neck to enable it to be filled. The bottle couldn't have a straight surface where the label was applied as this would create friction in the production line and damage the label. The product would also need to be distributed in the most efficient way, which dictated that the bottle comply with standards set by the crates and the vehicles that would transport it. In the market, the bottle would have to comply with standards for fridges and enable easy handling. The bottle would also have to fit vending machines and be suited to being dispensed by them. These are only a few of the constraints that may face a designer that need to be dealt with during the product design process.

The example above focused on a bottle. In all categories and product types there will be certain standards that have to be followed. In transport products, such as commuter trains, the requirements the designer will have to be aware of and comply with may seem endless. On top of all the usual standards governing the design of trains, the products need to accommodate the safety of passengers, ease of use and comfort. There are many other details that need to be considered in the design process. Paul Marchant, Head of Product Design at Transport for London, pointed out a problem with certain materials, such as stainless steel with a highly reflective finish: 'That does not work in station environments. People with visual impairments can find it very confusing. Its high reflectivity can create visual confusion as it mirrors surrounding light/colours and reduces contrast with objects around it' (Marchant 2011). This is just one of many concerns that need to be taken into account, and the level of detail gives an idea of the complexity of designing a train for public transport.

Every product that is designed, whether it be a water bottle or a commuter train, will have certain constraints that will have a massive effect on how the product looks and feels. In using the product as part of a brand-building strategy it is necessary to know the circumstances and different user scenarios the product will encounter, otherwise the product will most likely fail to build the brand as intended.

While establishing the limitations and constraints a product has to meet is a key consideration, at the same time it is important to design a unique product that has a character which differentiates it from other products on the shelves. Research into new technology and new manufacturing techniques often gives rise to new potential products. Constantly keeping up-to-date with such changes and opportunities makes it possible for the design team to design products for new user scenarios.

All Decisions Build the Brand

This chapter has explored the importance of informed decision-making in the design process when building a brand. Research is important in brand development, and is also a factor in making design a professional process. What kind of information designers rely on and how this informs the design process will define the product's role in building the brand. A company will need to respond to changes in the market context, but using only information from the market will result in the brand following rather than leading trends.

Establishing a direction in the design process is of key importance, but this direction can change based on insights that are gained during the process. The design process is perceived as being iterative, and will have key stages when important decisions need to be made. It is important to bear brand awareness in mind at all the design stages where decisions are made – both major and minor ones.

7

Innovation: Balancing Coherence and Change

Innovation is to take bold steps and walk in new directions. It is not an evolution and not a clear continuation of a line from the past. It is not a break with the past but doing something entirely different. On the Tripp Trapp chair we were a furniture producer that suddenly started making children's chairs. In reality we took a large step into a new segment. This is also something we did with the Xplory pushchair. We did not produce pushchairs and did not have the means of production for this. If the move is large enough, it is innovation to me.

Hilde Angelfoss, Director, Innovation & Design, Stokke

Innovation is essential to keep the brand and customer relations alive and give the company a key competitive advantage. In particular, this chapter will focus on some companies' main concerns when balancing a coherent identity with change so that the product remains relevant for the consumer.

Innovation in Product Design

WHAT IS INNOVATION IN BRAND-BUILDING?

Innovation is relevant to developing business models, processes, working methods or campaigns, but this book concentrates on product design and innovation. Both innovation and brand-building are recognised as key tools in developing a business. Nevertheless, they are not always seen as two sides of the same coin. Brand-building focuses on establishing coherence by communicating a message across a range of touch-points with the consumer. Innovation, on the other hand, has been associated with change, and even radical change in how people behave. This can create conflict if it is not managed skilfully.

The dilemma of change versus continuity has been a prominent issue for discussion in design circles since the early 1950s, when Raymond Loewy introduced the MAYA principle: 'Most Advanced, Yet Acceptable' (Loewy 2000: 277–83). This principle means that any new product that is designed should be as advanced as possible while remaining acceptable to the consumer. The MAYA principle is still important nowadays.

In this book, innovation is understood in terms of a much-quoted definition provided by Sir George Cox, former Director of the Victoria and Albert Museum in London: '"Innovation" is the successful exploitation of new ideas' (Cox 2005). This definition focuses on 'new ideas', but emphasises that these have to be exploited in some way. What makes such exploitation successful depends on the context. Success may consist of a change new ideas lead to, or in a business context it can be measured through profit margins. This chapter concentrates on successful brand-building – a new idea a product represents is only successful if it has a positive effect on how people think about the product or the company behind it. Therefore, innovation can serve as a tool to strengthen, reposition or maintain a brand.

In product development, the meaning of the term 'innovation' differs somewhat depending on the sector in which the company operates. A marketer at Ringnes (Carlsberg's Norwegian operation) used the term 'innovation' when a new flavour of bottled water was introduced successfully, while Hilde Angelfoss, Director for Innovation and Design at the Norwegian children's product company Stokke, used the term to refer to more fundamental changes. A new flavour of bottled water and something that changes people's behaviour are both examples of innovations, as they represent an idea that is exploited; the difference between the two lies in the degree of change that is created.

Until recently, the links between design, branding and innovation have not been explored in depth. Marketing researchers Michael Beverland, Julie Napoli and Francis Farrelly have proposed that the level of innovation determines the company's position in the market (Beverland et al. 2010). They introduced a model with two dimensions to structure brands. The first dimension represents the degree to which the company is driving the market as opposed to being market-driven. The second dimension represents the level of innovation, and whether the products launched are perceived as radical or incremental innovations. Their model provides insight into the role of innovation in a company's current market position. It will give the company an indication of where it is located in the market, and can help to raise its awareness of the perceived level of innovation in its competitors' products.

Another process has been developed by the Dutch designer and innovation strategist Eric Roscam Abbing: 'brand-driven innovation' (Abbing 2010). As the name suggests, brand-driven innovation describes how an innovation process can be driven by the brand. The process is divided into four phases. Phase one involves exploring brand usability – how people interact with and use the product. This process has similarities to the customer experience journey presented in Chapters 3 and 6. The second phase consists of developing an innovation strategy. The third phase concerns developing a design strategy. The fourth and final stage focuses on 'touch-point orchestration' – aligning all the company's touch-points with the consumer. By first developing an understanding of the current situation of the brand, then planning how to change, developing a design strategy and later implementing the change in a range of touch-points, the company will develop an innovation strategy that is driven by the brand.

TECHNOLOGY-DRIVEN INNOVATION

Innovation is traditionally associated with technology-based industries, where technology gives the impetus to develop new products. Technology is without doubt an important driver for innovation. When new technologies are invented, new products are developed in order to exploit them.

The innovation researcher and Harvard Business School professor Clayton M. Christensen uses the terms 'disruptive' and 'sustaining' to describe different levels of innovation in technology (Christensen 2003). He claims that while sustaining innovation increases product performance, disruptive innovation may result in worse product performance initially, and may even be more expensive. Even so, Christensen argues that companies need to keep investing in disruptive innovation because it leads to future competitive advantages.

New technology may not always lead to a disruptive experience or new user scenarios. The changes may not even be noticed by consumers. In the TV market, new technologies have recently been launched based on high-definition LED (light-emitting diode) and LCD (liquid crystal display) technologies. From the consumer's point of view the new LED TV sets provide significantly better quality, at least at first glance. Experts explain the difference in terms of an improved method of displaying black on TV screens. Nevertheless, many consumers may not find it obvious why there is a need to pay what at the moment is approximately twice as much for an LED TV compared to slightly less advanced technologies. In the near future the price will most likely fall as production volumes rise and production techniques improve, and the new technology may eventually be what everybody will seek when buying a new TV set. In cases like this the new technology will not create a disruptive impact on how the user experiences the product. However, over time it may lead to a disruptive change in the market if all consumers demand the new technology when buying a television.

For a long time the television industry has been strongly technology-driven, constantly exploring new user scenarios. This emphasis on technology also affects how televisions are presented to customers in stores. For non-expert customers this can provide an unsettling experience: they may not understand the technical data, and may be at the mercy of salespeople to explain the implications of specifications. The media market is rapidly moving from one-way broadcasting to interactive multimedia, where PCs, tablets and smart phones have become substitutes for the television. This indicates that the television industry is experiencing disruption and is about to undergo radical changes. In facing these changes a design approach to innovation will prove useful in developing new user scenarios for the television industry.

Currently there is an important debate in the technology industry about how open the process of innovation should be. Henry Chesbrough, Director at the Garwood Center for Corporate Innovation at the University of California, Berkeley, introduced the term 'open innovation' (Chesbrough 2006). He suggests that this represents a paradigm shift in how businesses should organise the innovation process. Instead of relying only on internal sources, the company should open up what have formerly been seen as the business's internal assets. In open innovation, how the business model is set up is of key importance. It is interesting to see how these open processes affect how the products and the brand are perceived. The innovation and design company IDEO is one organisation that has developed an open process it refers to as 'Open IDEO', in which it invites people to come together and be part of an open development process. The aim of this process is 'social good', and the initiative is run on a non-profit basis by IDEO. Participants are not rewarded financially but through a system where badges are allocated to the Internet profiles of key contributors (Open IDEO 2012).

The Danish toy manufacturer Lego is one of many companies that have used open innovation to build their brands (Open Innovators 2012). The Lego

Design byME service was established in 2005 (as 'Lego Factory'). It invited people to come together to design their own Lego characters. The complexity of delivering such a service meant that after six years the company decided to discontinue it. Lego said in a press release that it was committed to customisation and believed in the future of this approach, but Design byME had not lived up to Lego's quality standards. According to Lego, a total redesign of the service would have been necessary in order to continue it, and this would have made the products too expensive (Arnold 2011). This example illustrates the complexity and challenges established companies face in opening up part of their design process.

MAKING PROPOSALS TO PEOPLE

A design approach to innovation that is driven from a cultural perspective is presented by Roberto Verganti, Professor of Innovation Management at the Politecnico di Milano. Verganti's approach has its starting point in Italian brands such as Alessi and Artemide. Verganti has studied Italian design companies and their innovation processes, and states that there is a significant difference between the focus of a designer and that of an engineer in the product development process (Verganti 2010).

While engineers focus on innovation that is technology-driven, designers focus on innovation that is meaning-driven. His point is that product design can be understood in terms of how the product provides innovations in meaning. Design, he says, 'innovates meanings, and meanings make a difference in the market'. To what extent designers can innovate meaning is open to debate. Meaning in the context of design was described in Chapter 3. Meaning is how consumers interpret the product and what it represents for them. This is beyond the control of the designer or company. However, an important part of the design process is to picture the rational and emotional benefits a product will convey, and try to communicate it through the product. It is also important to design according to the company's own principles, and to express these principles through the form of the product. If the design intent is followed up by promotions or the company engages in activities that strengthen the intended message, it can be understood by the consumer as the meaning the product embeds.

One product described in Verganti's book *Design-driven Innovation* (2010) is the Italian company Artemide's Metamorfosi lamp. For customers, this changes the whole meaning of a lamp. From being simply a product to keep a room nice and bright, the lamp becomes a theatre in itself. Even though it is still a lamp, it adds a cultural dimension to the product. Another focus for Verganti's research is the Italian kitchen products company Alessi. Innovation, experimentation and helping people to improve their perception of the world is at the core of the company's mission. According to Verganti, what Alessi and Artemide have in common is that they do not base their strategies on looking at market changes and paying attention to what competitors are doing. In their strategies, and central to design-driven innovation, it is a question of making proposals to people. This strategy is not unlike Apple's approach. When Apple launched the iPhone in 2007 it made a proposal to people. People accepted this proposal, and Apple became a world leader in smart phones.

Building a Coherent Look and Feel

USING FACELIFTS TO REMAIN RELEVANT

Not every change in product design needs to be driven by a desire to innovate. Some changes are not a matter of introducing new ideas to the product, but rather making former ideas relevant as time goes by. Each period can be said to have its own look and feel, and most so-called 'timeless products' can be recognised as belonging to the period when they were designed when they are seen in relation to other products. To maintain a brand there is a need to change in minor ways just keep the product relevant in the current zeitgeist.

In the fast-moving consumer goods industries, incremental changes in product features or design references are made to give brands a facelift or in an effort to stay up-to-date with the current look and feel. On closer examination, even brands you may not think have changed will actually have altered a lot. However, because these changes have been very gradual, the product always feels relevant while retaining a strong connection to the brand identity. One FMCG product that is very well known for implementing successful incremental changes is Coca-Cola. Changes to the various elements of the form of the bottle – such as the grip, the silhouette and the space for the label – have been limited to minor facelifts over time. If you compare the first Coca-Cola bottle launched in 1916 (Vigo 2012) with the latest bottles, there have been significant changes in the form. The same can be said for its well-recognised logo. Changes such as facelifts are not perceived as innovations. They are incremental changes, but they do not involve introducing a new idea or concept in the product. These subtle changes are not always noticed by consumers, but they are important in updating the product's appearance. By constantly making minor changes to the bottle, the Coca-Cola Company manages to keep it up-to-date while at the same time referencing its past. Giving a product a facelift according to current trends and translating existing design cues into new objects enable companies to keep their brands relevant. Coupled with powerful commercials that explain why the brand is relevant today, they manage to build their reputation as 'brand icons' (Holt 2004).

Part of the design team's responsibility will be to design the product so that it remains relevant. In the design community there is an emphasis on radical innovation. A desire for drastic change could jeopardise the meaningful relationship a consumer has with the brand. The former Marketing Manager at Ringnes (Carlsberg's Norwegian operation) observed that many marketers seem too concerned that their brand is not seen as 'innovative and cool' (Michaelsen 2007). For some of Ringnes's brands, being 'cool' wasn't an issue – it was more important to be predictable. We as consumers buy certain products because of their predictability. For these brands, a strategy based on incremental innovation may be appropriate to maintain consumers' trust. Brands that are known for being very safe and reliable may find that previous incremental changes have secured this position for them. In such cases, a radical innovation process could lead to disturbance.

USING INCREMENTAL CHANGES TO REMAIN RELEVANT

Releasing products that do not succeed in the market could be a costly affair. Under Alfred P. Sloan's leadership in the 1920s, General Motors (GM) introduced three important concepts to reduce risks in product design: styling, annual release of new models, and 'dream cars'. The Vice President and Head of the Styling Department Harley Earl played an important role in developing these concepts (Sloan 1986).

Sloan's principle was that each of the cars should communicate a modern understanding of the GM style. The car should be in keeping with the times, and it should communicate what kind of brand the car belonged to: for example, a Buick should be distinct from a Chevrolet. The next car model in the same brand should also belong to a tradition reflected in the form of the car. According to this view, looking at cars over time there should be a gradual transition from one car to another. This technique allows changes in what the car represents, but at the same time maintains a style that people already recognise.

Using Chevrolet as an example, the first Chevrolet was launched in 1911 (Bayley 1990). This car incorporated references to the Ford Model T (which had been launched in 1908) and the horse-drawn wagon. From 1931 there was a more radical change in the appearance of cars and an increasing degree of streamlining. By 1954 the Chevrolet had received tailfins. This strategy of changing models annually had been defined in the late 1920s, and from 1911 to 1954 it is possible to observe a gradual change in the cars from model year to model year.

Harley Earl had a vision for the design of future cars. In the 1950s the car not only represented mobility, it *was* mobility, and mobility represented freedom (McCracken 2005). Bearing the optimism of the 1950s in mind, it is easy to see that cars from that era have references to aeroplanes and project the idea that the cars could fly. Using cars to represent visions of the future was a strong tool. These visions were developed as prototypes and presented to the public in car shows. These 'dream cars' also inspired the product design process. Historically, there have been great changes in the look and feel of the car, while there have been only gradual change from one annual car model to another.

By combining styling, annual launches of new car models and 'dream cars', car manufacturers can present their visions of the future. At the same time they can launch products that are within the bounds of current public taste. The use of 'dream cars' provides car manufacturers with the possibility of projecting the future without risking the disruptive leaps that would result if they introduced 'radical' car models straight to the market. This creates a dynamic within the brand. The future is projected through 'dream cars', while the cars that are launched on the market change only gradually, in tune with consumer preferences in the marketplace. The cars presented for sale include elements from the concept cars, but are not as radical.

CHANGING ESTABLISHED BRAND FEATURES

The look and feel of established brands can become old-fashioned, in which case incremental changes may not be enough to maintain the relationships the brand has represented in the past. Brands with well-established design references may in some cases have to make substantial changes to their design features in order to represent the same meaning. This is what the global fast food chain McDonald's experienced when a health and minimalism trend in society changed consumers' perception of its brand. Young people preferred to go to Starbucks instead of traditional fast food outlets. McDonald's did not see any option other than to radically change the meaning of its brand. It altered the product itself by introducing new salads on the menu, but it also had to change well-established design features that had hitherto communicated its brand. This has been of some assistance in repositioning its brand in the market.

In 2006 McDonald's launched a new exterior and interior redesign of its well-known outlets. For years they had featured a white plastic interior with a characteristic red roof. Bright shades of red, yellow and white dominated the experience. In 2006 McDonald's opted for a complete redesign. It decided to change the colour palette and move away from the familiar red it had established as its key colour over many years. John Miologos, McDonald's Vice-president of Worldwide Architecture, Design and Construction, told *BusinessWeek*: 'McDonald's promises to be a "forever young" brand,' and that the designers 'have to deliver on that promise' (Gogoi 2006).

McDonald's will always be about fast food, and critics may say that changes in the visual appearance of its outlets are not enough to change the perception of McDonald's as serving greasy food. On the other hand, there are signs that this might be about to change. Financial results also point to a significant increase in sales following the substantial change in the visual appearance of McDonald's outlets across Europe. These increases indicate that McDonald's is once again being understood as a young brand that appeals to its target audience (Werdigier 2007).

Repositioning the Brand through Products with Strong Stories

DEVELOPING A PRODUCT WITH A STRONG PRODUCT STORY

Developing breakthrough products has been recognised as an important factor in developing a company's revenue. This can create a 'blue ocean' for the company – a metaphor for launching products without having to compete with others for attention (Kim and Mauborgne 2005). This section will examine the case of Stokke as an example of this 'blue-ocean' strategy.

The Norwegian company Stokke has been successful in launching breakthrough products. A traditional wood furniture company, founded in 1932, its first really innovative product was a children's wooden chair – the Tripp Trapp, introduced on the market in 1972 (see Figure 7.1). The design process behind this chair was not driven by the desire to break the type form or establish a brand. It was driven by a desire to provide a solution to a need. The chair was designed by the Norwegian industrial designer Peter Opsvik in co-operation with Stokke. Opsvik works as a freelance designer, seeing his role as revealing needs that people have, rather than creating them. For him, an essential breakthrough in the design process for the Tripp Trapp was when he realised the need for a different type of chair. This was of greater importance to him personally than actually solving how to design the chair.

Opsvik realised that his son needed a new chair as he was growing out of a chair designed for smaller children (Opsvik 2007). In addition, existing children's chairs were low or had their own eating trays, which hindered interaction with the child when having a family meal. The idea was to develop a high chair that allowed children to be part of the family dinner table, and allowed the chair to 'grow' with the child by being easy to adjust to the child's height. He approached Stokke with this concept as they were open to new ideas from outside designers and already had experience in producing and distributing wooden furniture.

Introducing a product that created a new user scenario to the market allowed the child to become an active participant in the family dinner (Opsvik 2007). The product also broke with the existing type form – the typical form of the product (Molotch 2003). Product categories are important in defining what the product means to consumers. The use of categories has proved to be important in determining whether or not a product will be successful. The various product categories each have a type form which consumers understand as representing that specific product class. If it differs radically from the type form, the product will most likely break with its category.

Products that successfully manage to break the category rules and establish a new type form have an advantage over their competitors. If a radically different form is introduced, it may even establish a new category of products or a new, distinct type of product within a given category. How the product differs from what is typical in the category will then dictate perceptions of how 'radical', 'new' or 'innovative' the brand is.

As with many new inventions when they are launched in the market, the Tripp Trapp did not attract much interest from consumers until ten years after it was introduced. Today, forty years after its launch, many other companies have designed similar chairs. The Stokke Tripp Trapp chair is now seen as a

Figure 7.1   Stokke Tripp Trapp, designed by Peter Opsvik (1972)
Copyright Stokke AS.

typical example of what children's chairs should look like. Having ownership of this type form gives Stokke credibility in designing these chairs. The chair has also been an important revenue stream for the company. Stokke had the courage and patience to introduce a product with a radically different user scenario on the market. This has later been interpreted as meaningful, and the chair developed itself into a brand without any sophisticated branding strategy behind the chair.

The advertising industry has recognised that using breakthrough products could create so called 'baked-in' strategies. This is innovative products or business models with the marketing strategy embedded (Bogusky and Winsor 2009). A product with a radically different user scenario or look and feel can easily catch people's attention and be spread by word of mouth – Tripp Trapp is an example of this and Facebook is another. The Facebook service offered a user scenario that differed from earlier social networks, along with benefits for the user. An important aspect of this service involves allowing users to 'friend' people, which means they are self-motivated to publicise the service as a by-product of sending requests to their friends to join it. The marketing strategy is embedded in the service, and appears 'baked in'.

INNOVATION AS A TOOL TO CHANGE THE PERCEPTION OF THE BRAND

In 1997 the market situation had changed for Stokke. These changes threatened the sale of the Stokke Tripp Trapp. The chair had been a sales star for the company (Landmark 2007), but consumers preferred brands, not single products. In the first years of parenting, parents seek the comfort of well-known brands that have been developed with care for their newborn in mind. The fact that a chair was not the first object parents bought when they had a baby meant that they developed preferences for brands other than Stokke. At the same time the market had changed to relying on specialised retailers with the power to express their own demands. They were interested in selling products from companies that could deliver more than one product to their shops.

After a strategic decision process, Stokke decided to make a drastic move and split the company into two divisions, which later became two separate companies. One part concentrated on wooden furniture, and the other on children's products. The Stokke brand was applied to the children's products, thus changing from signifying wooden furniture alone (the most successful piece of furniture already being a children's chair). Since Stokke was a product-driven company, this change began with altering the product portfolio before changing the communication strategy.

An important part of this change was to establish the company's strategic direction. To develop this direction, Kristine Landmark, Stokke's Managing Director, invited a group of staff to a workshop. They asked themselves questions such as: 'What is it that makes the Tripp Trapp desirable from a consumer perspective? What is it that makes it unique?' In 1997 they had not developed a brand strategy for the Stokke Tripp Trapp chair, but it was widely recognised as a chair for children. The answer to the questions was that this product had a human-centred purpose – it was developed with a focus on the child's wellbeing and need to belong to the family. The facts that the chair grew with the child and that it was made of wood meant the design was sustainable. The chair also gave the child the opportunity to be active and feel comfortable, which led to satisfied parents.

The company had only few products in the children's products category – the chair and a few pillows. To be perceived as a children's product brand, it needed to expand its portfolio. The next step for the company was to develop other products in a category it called 'care' –products for domestic use, such as beds and changing tables. In developing these products, the idea of 'growing with the child' was still central to the product design process (Refsum 2007). Stokke had knowledge of furniture products, so it was not too challenging for it to develop them. Within the children's category, the transport segment is a crucial sub-category. Transport products were recognised by Stokke as important in developing a complete children's brand, but they were also important because of the mobility the products implied. When Stokke decided to move into this category, its aim was not to develop a pushchair, but to develop a product in the transport segment (Refsum 2007). This gave Stokke greater freedom, as it did not look specifically at the category of pushchairs first, but started with a broader view of children's transportation needs. One area that seemed to hold potential for a Stokke product was transportation in an urban environment.

Stokke gave five design agencies a brief which invited them to come up with a product that was the ultimate design for an urban environment. The design company K8 won the competition with a radical new idea that was inspired by the Tripp Trapp. The concept proposed was a stick-like chassis with flexible seating positions, allowing children to explore the world and interact with their parents on the move.

When Stokke Xplory was launched on the market there was a distinct type form in the pushchair category. By focusing on children's needs in the product development process and maintaining its strong desire not to fall into the 'me too' trap, Stokke managed to develop a product with a strong character that appeared radically different from all other pushchairs on the market (see Figure 7.2). When the product was launched it was such a successful break with the category that *Time Magazine* named the Stokke Xplory one of the most innovative products released in 2004 (Hamilton et al. 2004).

Figure 7.2   Stokke Xplory, Version 1 (2003)
Copyright Stokke AS.

With this pushchair Stokke managed to create a new segment within the pushchair category. The Stokke Xplory was also a successful innovation in terms of branding, helping the company to change perceptions of the Stokke brand. One of the key learning points from the marketing of the Stokke Xplory was that the innovative design of the product challenged user expectations, so it was to some extent necessary to teach buyers how to use it. The story of the pushchair needed to be followed up in the marketing material, which meant that the product played an integral role in the marketing campaign, rather than adding a story at the end that had little to do with it.

Developing products that disrupt a particular category can be a challenge when building a brand. In branding, coherence is important in order to be seen as part of a brand. A radical change could jeopardise recognition. However, in the case of Stokke, a break with the past as a furniture producer was necessary. In this scenario the attention the pushchair gave the company was crucial in establishing a new direction for the brand.

For Stokke, innovation is important, but being innovative is not a goal in itself. Innovation is a means to create products that offer new user scenarios or added value to users. According to Kristine Landmark (2007), innovation means that 'when [Stokke] makes basic products such as Xplory and the Tripp Trapp, these should be innovative. This innovation should be a benefit for the child. It should be in the best interest of the child.' Stokke's innovation strategy focused on the wellbeing of children. This focus also provided the potential to convey a coherent message about the brand.

Stokke has found it necessary to combine incremental product changes and innovation with more substantial changes in order to build a coherent brand that remains relevant (see Figure 7.3). Since the first launch, Stokke has released a number of new versions of the Xplory on the market. By balancing an initial substantial change in the product with later incremental changes, the company has secured a return on its investments. This also leads to users gaining access to the best possible solution for a product within the given category.

Figure 7.3   Stokke Xplory, Version 3 (2010)
Copyright Stokke AS.

DESIGN-DRIVEN INNOVATION VERSUS STRATEGIC-DRIVEN INNOVATION

The examples of Alessi and Stokke illustrate two different approaches to using product design as an important factor in building a brand. When considering innovative product design as part of a branding strategy, it is important to understand that each product will be perceived in a certain way. The marketing department will have to follow this up to build a coherent brand across a range of products (Angelfoss 2007).

The companies Roberto Verganti describes are strong brands. Both Alessi (kitchenware) and Artemide (lighting) have managed to build strong portfolios of products that are significantly different to others on the market. At the core of these brands is a desire to produce a specific type of product that conveys a cultural message. Using so-called 'design heroes' (Philippe Starck, Michael Graves, Karim Rashid and others) to communicate the story of a product works well in the domestic category, and this strategy means that domestic objects from these companies also serve as art objects.

Stokke built its reputation on wooden furniture and using individual designers' philosophies to design new products. For Stokke, it was not viable to continue with this design approach when moving into the children's products segment, so it changed from an innovation strategy that was led by individual designers to a strategic-driven innovation approach (Angelfoss 2007). In this new strategic-driven approach the Stokke design project started by revealing a need within the transport category. A user-centred need was identified initially, and the result of the process was a product that was different from other products on the market. This happened because the idea was not to design a pushchair, but to identify a need in the transport segment for which Stokke wanted to define a solution.

Design-driven innovation, based on 'design heroes' and the strategic-driven design approach, as outlined in the Stokke case, are equally important in building strong brands. The choice of design approach will depend on the context within which the company functions, including the market situation. It is important to develop a strategy that sets out how design is to be used to help to build the brand. In this strategy, the reason for innovation will need to be identified and communicated clearly, both internally and externally.

PLANNING CHANGES

Both incremental and more radical changes of established product features and references will play important roles in building the brand. Incremental changes are important when the consumer has an emotional attachment to a product that is already on the market. When the product references are well-established, the product may just need a facelift to stay up-to-date and follow relevant trends. However, if innovation is part of the brand DNA, the company will constantly need to push itself to launch new products that are innovative.

To keep track of changes and maintain coherence, it is best to create an innovation architecture that embodies information about what the right changes are for a particular brand (Skarzynski and Gibson 2008). An innovation architecture is a system that explains the relationships between the different product innovations the company has developed. This can be helpful in keeping track of different products and their level of innovation and consistency. An innovation architecture allows the company to keep track of innovation and how much its products have changed, helping designers to understand how design features contribute to building the look and feel of the brand, and also how trends have influenced the product.

Useful questions to ask when developing the innovation architecture include:

- How does the brand DNA relate to the product?

- What is the balance between incremental changes and leaps?

- What position does the product occupy in the brand architecture?

- What is the individual product's role within the product category?

- How has the customer understood the product?

- What is the historical meaning of the product?

- Which references should be built into the product?

- Which trends does the product comply with?

If the brand consists of just one type of product, it should be easier to keep track of innovations. However, many brands include a range of products and sub-products that have developed over time. Some global companies may even find that they have lost control of their brand because too many variants of the same products have been released.

Substantial changes or incremental approaches will not have the same effect for every brand. The context determines whether a new product will be perceived as an innovation. When considering this, there are three contexts a product needs to relate to: the company's heritage and product portfolio; the market and the category the product belongs to, and the society including the user scenario the product addresses. If the product's context changes, the product may also be perceived differently. The company needs to understand the particular cultural context it is part of, rather than merely focusing on what its competitors are doing in the marketplace.

The frequency of changes in products varies across categories. In categories where technology is a driver of innovation, such as consumer electronics, products can soon look outdated because of the high tempo of the market. The mobile phone category used to see many minor changes. As mentioned by Cathrine Movold in Chapter 4, it was a race, with companies competing over who could offer the most advanced mobile phone features while developing smaller and smaller handsets. When Apple went in a completely different direction with the iPhone in 2007, the market experienced a change. The iPhone was considerably larger than other handsets, but allowed the user to interact with the phone in a different way. When the iPhone was released, almost overnight many mobile phones were seen as outdated, despite their new functional features.

An important aspect of the industrial design process is to plan how the product will relate to these changes. Designer, researcher and innovation strategist Joanna Brassett worked for some of the major design companies in the United Kingdom before founding her own company. According to Brassett, product development timelines are useful tools for her design team, helping it to define how far new concepts can go. They are useful in developing the necessary coherence and for brand recognition, as well as forecasting how cultural changes will affect the perception of the product as being innovative or not (Brassett 2011). The design team can therefore arrive at an idea of what will be acceptable within the current social and cultural context while developing a sense of what may be acceptable in the future. Simply mapping out all the concepts and possible changes along timelines enables the design team to gain a better understanding of the appropriate level of innovation in a product it is about to develop.

The Brand as the Driver of the Process

This chapter has considered how product innovation can serve as an important tool in building a brand. To qualify as an innovation, a new invention needs to successfully exploit new ideas. How an innovation succeeds in building meaningful relations between people and the brand is of key importance in brand-building.

When using product design innovation to build a brand, it is important is to be aware of changes that will have an effect on how people perceive the brand, and in particular whether the innovation will disrupt the user experience. Finding the right balance between incremental changes and more disruptive ones is of key importance in harnessing innovation as a tool to build a brand. This is not a one-off task, but a dynamic learning process. The company therefore needs to listen constantly to society to understand what level of innovation will be appropriate for its product portfolio.

In innovation, the design approach can focus on the user scenario or revealing needs, and it can also focus on developing products that have a cultural reference or embedded philosophy that becomes important for the user. A design approach to the product innovation process can be successful in building a brand. It is important for the company to identify a narrative that is conveyed through its products in order to build a coherent brand, even though its products may appear different from one another. In using innovation as an important part of its branding strategy, the company will need to define its reason for designing new products. The answer to this question may provide the narrative that becomes the brand.

Epilogue

The Use of Design to Build the Brand: A Strategic Question

The aim of this book has been to examine how product design can be integrated into a company's brand-building activities. Starting with the question of how branding can inform the design process, I first tried to find a model or a theory of how to transform the intangible brand into products. However, when I embarked on my study I soon realised that this was a tactical approach to a strategic question. Along the journey I have talked with many designers, all of whom had good ideas about how to express and develop the brand through the product. However, they needed to have identified a relevant direction, and more importantly, the company environment needed to promote design. The organisation's leadership must have an appreciation for design in order to enable the changes required.

The basic principle in developing a product that builds a brand is first to identify a story or a direction that is relevant for users, then to design the product within this context. As soon as this direction has been established, it can guide the various stakeholders involved in the product design process through the journey of making design decisions. Throughout this journey the team will have to be open to changes as it learns about the product being developed. The real learning about the product's role in building the brand will come when the product is released on the market and customers start to engage with it.

Each of the chapters in this book provides guidelines, principles or models of how to use the design process when building a brand. The key points are summarised in Table E.1.

Table E.1   Summary of key concerns

Where	What	Description
Chapter 1	Define the strategy and role the product should play in the brand strategy.	How design helps to build the brand is an important question in the company's strategy. In defining the strategy, the company needs to consider how involved the product is in telling the story about the brand, and how engaged all the stakeholders are in telling this story.
Chapter 2	Define the values that form the basis for (internal and external) engagement with the brand.	Branding and product design are cultural processes driven by values. The starting point is to design these values and the principles the company would like to engage with. The next stage is to engage with consumers. As building the brand is a process that involves all stakeholders, dynamic interplay with them is what shapes the brand.
Chapter 3	Identify the product DNA and a story about how the product serves to build the brand experience.	In strong brands the product is an integrated aspect of building the brand. Coherence plays a major role in this. Over time the company should seek to define its product DNA.
Chapter 4	Define the Brand Activity Framework. Understand the larger context for the product design process, both internally in the company and as part of the market and cultural context – align touch-points.	The product design process is part of a dynamic interplay between internal and external inputs. This interplay shapes both what the product is about and what the brand is about.
Chapter 5	Develop a culture that maintains a knowledge base about the brand and the product.	Each design process plays its part in building the heritage of the brand. These processes are assets for the company that should be maintained and made available to the design team. Knowledge of past projects is important, but a more important factor is deciding who should design the products that build the brand.
Chapter 6	Identify key research input and define a direction for the design process.	Informed decision-making is crucial in defining how the product should help to build the brand. The design team will have to be selective in how input from the market and society is used as this is an important consideration in defining how the product can help to build the brand.
Chapter 7	Identify a balance between coherence and change that strikes the right tempo for the brand.	One of the key questions in using the design process as part of brand-building is how much the next product should change. Innovation in terms of branding is successful when the new product serves to build the brand in a way that is beneficial.

Product Design Plays an Important Role in the Discussion with Society

The dynamic nature of the branding and design processes means that the principles that define them are constantly changing. Both design and branding follow trends that will be vociferously argued for over a certain period, but will then change. However, some key principles are more constant, so they will be helpful in the future.

For designers, both modernism and post-modernism have been important factors in defining principles that still work nowadays. In modernism, a lasting tenet is the principle of being led by an idea that inspires the design process to arrive at the final details and finish of the product. From post-modernism, it is valuable to bear in mind the cultural relevance of the products and brands. An important point is that no single idea or set of values or principles defined by designers will drive the design process in every company. For a time the value-set in a company will need to be shared, but these values will need to change gradually to adapt to new circumstances, both internal and external.

There have been changes in general management principles that have parallels with those in design and branding. The concept of 'shared value' created by Michael Porter and Mark Kramer at Harvard University was mentioned in the Introduction. This concept entails a change in businesses' attitudes towards their corporate social responsibility, from being a 'feelgood' add-on the company undertakes as a promotional activity to acknowledging that the company can conduct value-driven activities that provide a profit while serving a greater good (Porter and Kramer 2011).

Shared value therefore includes a modified view of companies' role in society – something that is beyond profit, and where branding plays an important role. The brand is the impression someone has of a company. All the activities the company undertakes and how they engage with different stakeholders are understood as building the brand. The product design process can play an important role in this. Product design becomes an important mediator for the company. Companies have an opportunity to realise their visions for the future through the design of their products.

References

Aaker, D. 1996. *Building Strong Brands*. London: Simon & Schuster.
—— and Joachimsthaler, E. 2000. *Brand Leadership*. New York: Free Press.
Aaker, J. 1997. 'Dimension of brand personality', *Journal of Marketing Research*, XXXIV, 347–56.
Abbing, E.R. 2010. *Brand-driven Innovation: Strategies for Development and Design*. London: Ava Publishing.
Alessi. 2012. *Alessi Shortly: The Mission* (online: Alessi). Available at: http://www.alessi.com/en/company/briefly-alessi (accessed: 29 February 2012).
Anderson, C. 2009. *Free Economy: The Future of Radical Price*. New York: Hyperion Books.
AppleInsider. 2010. 'Apple now largest mobile device company in the world' (online, 27 January). Available at: http://www.appleinsider.com/articles/10/01/27/apple_now_largest_mobile_device_company_in_the_world.html (accessed: 12 February 2012).
Arnold, A. 2011. 'Design byME to close in January' (online: The Lego Group, 9 November). Available at: http://aboutus.lego.com/en-gb/news-room/2011/november/design-byme-to-close-in-january (accessed: 1 July 2012).
Arnold, B. 2006. 'Jay-Z pours away the Cristal', *The Guardian*, 16 June.
Arup, O.N. 1970. 'Ove Arup's Key Speech' (online: Ove Arup & Partners, 9 July). Available at: http://www.arup.com/Publications/The_Key_Speech.aspx (accessed: 22 July 2012).
Autocar. 2007. 'Autocar re-invents the Routemaster' (online: *Autocar*, December). Availableat:http://www.autocar.co.uk/car-news/motoring/autocar-re-invents-routemaster (accessed: 22 July 2012).
Barnett, E. 2011. 'Amazon Kindle ebooks outsell paperbacks', *Daily Telegraph*, 28 January.
Bayley, S. 1990. *Harley Earl*. New York: Taplinger Publishing.
Beckwith, D. 2004. 'Design's strategic role at Herman Miller – an interview with Don Goeman', *Design Management Review*, 15(2), 40–45.

Beverland, M. 2009. *Building Brand Authenticity: 7 Habits of Iconic Brands*. Basingstoke: Palgrave Macmillan.

——, Napoli, J. and Farrelly, F. 2010. 'Can all brands innovate in the same way? A typology of brand position and innovation effort', *Journal of Product Innovation Management*, 27, 33–48.

Blindheim, T. 2004. 'Forbruk som lyst og nytelse', in T. Blindheim et al. (eds), *Forbruk: Lyst, makt, iscenesettelse eller mening?* Oslo: Cappelen Akademisk Forlag, 27–80.

Bogusky, A. and Winsor, J. 2009. *Baked In*. Evanston, IL: Agate Publishing.

Brown, T. 2008. 'Design thinking', *Harvard Business Review*, June, 84–92.

Burkhardt, F. and Franksen, I. 1980. *Design: Dieter Rams*. Berlin (West): Gerhardt Verlag.

Chandler, D. 2007. *Semiotics: The Basics*, 2nd edn. New York: Routledge.

Chesbrough, H. 2006. *Open Innovation: The New Imperative for Creating and Profiting from Technology*. Boston, MA: Harvard Business Press.

Christensen, C. 2003. *The Innovator's Dilemma: The Revolutionary Book that Will Change the Way You Do Business*. New York: Harper Business Essentials.

The Coca-Cola Company. 2010. 'I LOHAS Mikan – a new flavored water from the I LOHAS brand' (online: The Coca-Cola Company, 3 June). Available at: http://www.thecoca-colacompany.com/dynamic/press_center/2010/06/i-lohas-mikan----a-new-flavored-water-from-the-i-lohas-brand.html (accessed: 5 August 2011).

Cooper, R. and Press, M. 2003. *The Design Experience: The Role of Design and Designers in the Twenty-first Century*. Aldershot: Ashgate.

Cox, G. 2005. *Cox Review of Creativity in Business: Building on the UK's Strengths*. London: H.M. Treasury.

Dempsey, A. 2002. *Styles, Schools and Movements: The Essential Encyclopaedic Guide to Modern Art*. London: Thames & Hudson.

Design Museum. 2011. 'Memphis: Product + furniture designers (1981–1985)' (online: Design Museum). Available at: http://designmuseum.org/design/memphis (accessed: 16 May 2011).

The Economist. 2001. 'Special report: Brands – Who's wearing the trousers?', *The Economist*, 6 September, 28–30.

——. 2006. 'Bubbles and bling', *The Economist*, 8 May.

——. 2010. 'Middle Kingdom meets Magic Kingdom', *The Economist*, 28 August.

——. 2012. 'The Ideas Economy' (online: *The Economist*). Available at: http://www.economist.com/events-conferences/americas/ (accessed: 20 October 2012).

Florida, R. 2002. *The Rise of the Creative Class and How it's Transforming Work, Leisure, Community and Everyday Life*. New York: Perseus Book Group.

Ford, H. 1923. *My Life and Work*. London: William Heinemann.

Friedman, M. 1970. 'The social responsibility of business is to increase its profits', *The New York Times Magazine*, 13 September.

Geek&Poke. 2010. 'The "Free" Model' (online, 21 December). Available at: http://geekandpoke.typepad.com/geekandpoke/2010/12/the-free-model. html (accessed: 25 February 2012).

Glancey, J. 2010. 'The new Routemaster bus is a design cacophony', *The Guardian*, 11 November.

Gobé, M. 2001. *Emotional Branding: The New Paradigm for Connecting Brands to People*. New York: Allworth Press.

Godiva. 2011. *History of Godiva* (online: Godiva Chocolatier Inc.). Available at: http://www.godiva.com/about/faq.aspx (accessed: 16 May 2011).

Gogoi, P. 2006. 'Mickey D's McMakeover: The heavy plastic look is history: A clean, simple design is on the way in', *BusinessWeek*, 14 May.

Graeber, D. 2001. *Toward an Anthropological Theory of Value: The False Coin of Our Dreams*. New York: Palgrave.

Green, O. and Rewse-Davis, J. 1995. *Designed for London: 150 Years of Transport Design*. London: Laurence King Publishing.

Hamilton, A.M. et al. 2004. 'Coolest Invention 2004: Kid friendly', *Time Magazine*, 29 November.

Heskett, J. 2002. *Toothpicks & Logos: Design in Everyday Life*. Oxford: Oxford University Press.

Hestad, M. 2007. 'Pure shape – to realise intended meaning in practise', in S. Kyffin, L. Feijs and R. Young (eds), *Design and Semantics of Form and Movement: DeSForm 2007*. Newcastle: Northumbria University.

——. 2008. *Den kommersielle formen: Om merkevarekonteksten som utfordring for industridesignernes behandling av form*. Oslo: Arkitektur- og designhøgskolen i Oslo.

Holt, D. 2002. 'Why do brands cause trouble? A dialectical theory of consumer culture and branding', *Journal of Consumer Research*, 29 (June), 70–90.

——. 2004. *How Brands Become Icons: The Principles of Cultural Branding*. Boston, MA: Harvard Business Press.

I Lohas. 2011. 'I Lohas' (online: Coca-Cola (Japan) Company Ltd). Available at: http://i-lohas.jp (accessed: 16 May 2011).

Ind, N. and Bjerke, R. 2006. *Branding Governance: A Participatory Approach to the Brand Building Process*. Oslo: Norli.

Interbrand. 2011. *Best Global Brands 2011*. London: Interbrand.

Isaacson, W. 2011. *Steve Jobs*. London: Little, Brown.

Jevnaker, B. 1995. *Den skjulte formuen: Industridesign som kreativ konkurransefaktor, SNF-rapport*. Bergen: Stiftelsen for samfunns- og næringslivforskning.

Johnson, B. 2008. *Getting Londoners Moving: Boris Johnson Transport Manifesto* (online: *The Guardian*). Available at: http://image.guardian.co.uk/sys-files/Guardian/documents/2009/04/27/Transportmanifesto.pdf (accessed: 22 July 2012).

Jones, P. 2006. *Ove Arup, Master Builder of the Twentieth Century*. New Haven, CT: Yale University Press.

J.P. Chenet. 2011. 'J.P. Chenet: History' (online: Grands Chais de France) Available at: http://www.jpchenet.com/historique-en.html (accessed: 5 August 2011).

Julier, G. 2000. *The Culture of Design*. London: Sage.

Karjalainen, T.-M. 2004. *Semantic Transformation in Design: Communicating Strategic Brand Identity through Product Design References*. Helsinki: University of Art and Design.

Karjalainen, T.-M. and Snelders, D. 2010. 'Designing visual recognition for the brand', *Journal of Product Innovation Management*, 27 (Special Issue).

Keferl, M. 2009. 'Bottle innovation: I LOHAS from Coca-Cola twists for the environment' (online: Japan Trends, 3 June). Available at: http://www.japantrends.com/bottle-innovation-i-lohas-from-coca-cola-twists-for-the-environment (accessed: 5 August 2011).

Kifner, J. 2002. 'Public Lives: Where ice water is an insult, and tap is a disgrace', *The New York Times*, 14 February.

Kim, W.C. and Mauborgne, R. 2005. *Blue Ocean Strategy*. Boston, MA: Harvard Business Press.

Klein, N. 2001. *No Logo*. New York: Flamingo.

Leadbeater, C. 2009. *We-think: Mass Innovation Not Mass Production*, 2nd edn. London: Profile Books.

Leo Burnett. 2009. 'Leo Burnett Human Kind Video' (online: YouTube). Available at: http://www.youtube.com/watch?v=-OpYn-nlSWo (accessed: 22 July 2012).

Li, C. and Bernoff, J. 2008. *Groundswell: Winning in a World Transformed by Social Technologies*. Boston, MA: Harvard Business Press.

Lindstrom, M. 2005. *Brand Sense: How to Build Powerful Brands through Touch, Taste, Smell, Sight and Sound*. London: Kogan Page.

Lockwood, T. 2010. *Design Thinking: Integrating Innovation, Customer Experiences, and Brand Value*. New York: Allworth Press.

Loewy, R. 2000. *Industrial Design*. London: Laurence King.

The Londonist. 2011. 'In pictures: The new bus for London' (online: *The Londonist*, 16 December). Available at: http://londonist.com/2011/12/in-pictures-the-new-bus-for-london.php (accessed: 29 February 2012).

Louis Vuitton. 2011. 'Louis Vuitton Women Fall/Winter 2010–11 Live Fashion Show' (online: Facebook/Louis Vuitton) Available at: http://www.facebook.com/event.php?eid=335718777125 (accessed: 16 May 2011).

Mark, M. and Pearson, C.S. 2001. *The Hero and the Outlaw: Building Extraordinary Brands through the Power of Archetypes*. New York: McGraw-Hill.

Martin, R. 2009. *The Design of Business: Why Design Thinking is the Next Competitive Advantage*. Boston, MA: Harvard Business Press.

McCracken, G. 1990. *Culture & Consumption*. Bloomington, IN: Indiana University Press.

——. 2005. *Culture & Consumption II: Markets, Meaning, and Brand Management*. Bloomington, IN: Indiana University Press.

McKendrick, N., Brewer, J. and Plumb, J.H. 1982. *The Birth of a Consumer Society: The Commercialization of Eighteenth-century England*. London: Europa Publications.

McKeown, M. 2012. *The Strategy Book*. London: FT Publishing.

Michl, J. 1995. 'Form follows WHAT? The modernist notion of function as a *carte blanche*', *Magazine of the Faculty of Architecture & Town Planning*, Winter (10), 31–20.

——. 2007. 'A case against the modernist regime in design education', revised version of paper presented at CUMULUS (The International Association of Universities and Colleges of Art, Design and Media), Bratislava, Slovakia, 12 October 2007. Available at: http://janmichl.com/eng.apartheid.html (accessed: 10 January 2012).

Miller, S. 2010. 'Do You Have the Time' (online: Disney Parks Blog). Available at: http://disneyparks.disney.go.com/blog/2010/11/do-you-have-the-time-mickey-mouse-watches-at-disney-parks/ (accessed: 25 November 2012).

Molotch, H.L. 2003. *Where Stuff Comes From: How Toasters, Toilets, Cars, Computers, and Many Other Things Come to Be as They Are*. New York: Routledge.

Morris, B. 1976. 'A Designer Gets Ahead of Himself', *The New York Times*, 21 September, 56.

Neumeier, M. 2009. *The Designful Company: How to Build a Culture of Nonstop Innovation*. Berkeley, CA: Peachpit Press.

Norman, D. 1998. *The Design of Everyday Things*. New York: MIT Press.

——. 2009. 'Technology first, needs last: The research–product gulf', *Interactions*, 17(2).

Nussbaum, B. 2009. 'Technology vs. design – what is the source of innovation?' *BusinessWeek*, 1 December.

Open IDEO. 2012. (online) Available at: http://www.openideo.com (accessed: 25 November 2012).

Open Innovators. 2012. 'List of open innovation & crowdsourcing examples – best practices' (online: Open Innovators). Available at: http://www.openinnovators.net/list-open-innovation-crowdsourcing-examples/ (accessed 22 July 2012).

Paulsson, G. 1919. *Vackrare vardagsvara*. Stockholm: Svenska slöjdföreningen.

Porter, M.E. 1996. 'What is strategy?', *Harvard Business Review*, November–December, 61–78.

—— and Kramer, M.R. 2006. 'Strategy and society: The link between competitive advantage and corporate social responsibility', *Harvard Business Review*, December.

—— and Kramer, M.R. 2011. 'Creating shared value', *Harvard Business Review*, January–February.

Pressman, G. and Kerner, N. 2007. *Chasing Cool: Standing Out in Today's Cluttered Marketplace*. New York: Atria Books.

Pye, D. 1978. *The Nature and Art of Workmanship*. Cambridge: Cambridge University Press.

Quinn, T. (Executive Producer). 2011. *Steve Jobs: Billion Dollar Hippy* (television broadcast, 14 December). London: BBC Two.

Roberts, K. 2004. *Lovemarks: The Future Beyond Brands*. New York: Powerhouse Books.

Rockwell, C. 2010. 'The mathematics of brand satisfaction', *Design Management Review*, 19(2), 221–9.

Schön, D. 1983 [2011]. *The Reflective Practitioner*. Farnham: Ashgate.

Skarzynski, P. and Gibson, R. 2008. *Innovation to the Core: A Blueprint for Transforming the Way Your Company Innovates*. Boston, MA: Harvard Business Press.

Skjerven, A. 2003. 'Great expectations – the foundation of a design concept', in W. Halén and K. Wickman (eds), *Scandinavian Design Beyond the Myth*. Stockholm: Arvinius Förlag.

Sloan, A.P. 1986. *My Years with General Motors*. London: Sidgwick & Jackson.

Snelders, D. and Lloyd, P. 2003. 'What was Philippe Starck thinking of?', *Design Studies*, 24, 237–53.

Stompff, G. 2008. 'Embedded brand: The soul of product development', *Design Management Review*, Spring, 38–46.

TfL. 2012. 'New bus for London' (online: Transport for London). Available at: http://www.tfl.gov.uk/corporate/projectsandschemes/15493.aspx (accessed 29 February 2012).

Thjømøe, H.M. 2003. 'Conceptual Models for Defining, Influencing and Communicating the Brand: What Do We Really Know?', paper presented at *The 2nd International Conference on Research in Advertising*. Amsterdam: University of Amsterdam.

Twitchell, J.B. 2001. *Living It Up: Our Love Affair with Luxury*. New York: Colombia University Press.

Valtonen, A. 2005. 'Six Decades – and Six Different Roles for the Industrial Designer', paper presented at the Nordes Conference *In the Making*, Copenhagen, Denmark, 30–31 May 2005.

———. 2007. *Redefining Industrial Design: Changes in the Design Practice in Finland*, Publication Series of the University of Art and Design Helsinki A – 74. Helsinki: University of Art and Design.

Verganti, R. 2010. *Design-driven Innovation*. Boston, MA: Harvard Business Press.

Vigo. 2012. *Earl R. Dean Collection* (online: Vigo County Public Library). Available at: http://www.vigo.lib.in.us/archives/inventories/business/dean.php (accessed: 29 February 2012).

Vihma, S. 1995. *Products as Representations: A Semiotic and Aesthetic Study of Design Products*, Publication Series of the University of Art and Design Helsinki UIAH A – 14. Helsinki: University of Art and Design Helsinki UIAH.

Vogel, C.M. 2010. 'Notes on the evolution of design thinking: A work in progress', in T. Lockwood (ed.), *Design Thinking: Integrating Innovation, Customer Experiences, and Brand Value*. New York: Allworth Press, 3–14.

Vossoughi, S. 2007. 'The best strategy is the right strategy', *Design Management Review*, 18(4).

Walker, R. 2009. *Buying In: What We Buy and Who We Are*. New York: Random House.

Weckström, C. 2011. 'Designer Profile: Global Head of Consumer Experiences at Lego'. MA Industrial Design, Central Saint Martins College of Arts and Design, London, 22 November.

Wenger, E. 2006. 'Communities of practice: A brief introduction' (online: Etienne Wenger, June). Available at: http://www.ewenger.com/theory/communities_of_practice_intro.htm (accessed: 22 July 2012).

Werdigier, J. 2007. 'For McDonald's, a European redesign starts to pay off', *The New York Times*, 17 August.

Wheeler, A. 2006. *Designing Brand Identity*. Hoboken, NJ: John Wiley & Sons.

Zaltman, G. 2003. *How Customers Think: Essential Insights into the Mind of the Market*. Boston, MA: Harvard Business School Press.

Interviews

Angelfoss, Hilde. 2007. Product Development Manager at Stokke AS, Håhjem, Ålesund, 27 August.

Brassett, Joanna. 2011. Founder and Director at INTO Ltd, London, 21 December.

Gran, Martin. 2007. Consultant at Scandinavian Design Group, Oslo, 23 April.

Hellerud, Geir. 2007. Product Development Manager at Jordan AS, Oslo, 14 September.

Horsrud, Hanan. 2007. Marketing Manager at Ringnes AS, Oslo, 25 May.

Landmark, Kristine. 2007. Director at Stokke AS, Ålesund, 28 August.

Marchant, Paul. 2011. Head of Product Design at Transport for London, London, 13 December.

McCullagh, Kevin. 2012. Founder and Director at Plan Strategic Ltd, London, 18 January.

Michaelsen, Oscar. 2007. Former Head of Marketing at Ringnes AS, Oslo, 31 May.

Movold, Cathrine. 2011. Industrial Designer at Making Waves AS, Oslo/London, 21 December.

Mycroft, Damian. 2011. Head of Industrial Design, Hewlett-Packard Company, London, 16 June.

Opsvik, Peter. 2007. Industrial Designer, Oslo, 5 September.

Øxseth, Geir. 2007. Industrial Designer, Oslo, 21 October.

Refsum, Bjørn. 2007. Former Project Manager at Stokke AS, Ålesund, 10 October.

Wentworth, Michelle. 2007. Category Development Manager at Jordan AS, Oslo, 14 September.

Index